Prologue

Every part of our body needs regular healthy nutrition and exercise, otherwise, it starts to lose power, and we will gradually die before our time.

As we grow older, different parts of our body get weaker and we encounter a variety of health issues. For example, our vision normally begins to weaken as we hit our forties, but some people may face it even earlier. There are also people who do not encounter any vision trouble even after the age of fifty.

There are a lot of factors that can affect vision, but our aim is not these issues. We, therefore, leave this matter to physicians and experts, but obviously, if you are a sportsman or doing some regular exercise with a healthy diet, the possibility of early weakness in your body is much lower than other people.

The brain is one of the most important parts of the human body, as it controls the functionality of other parts; therefore, it needs special care and attention. Proper diet and regular brain exercise keep our brain cells healthy and cognitive functionality strong.

In order to keep the brain fresh and sharp, we need to challenge it regularly by doing the mental exercise. Games such as chess, crossword puzzles are light and easy mental exercises, but doing mathematics and science are also very helpful.

Not everyone is comfortable with mathematics and science; therefore, to avoid hating mathematics and inversely loving it, I am introducing a new method and trick to do basic mathematics, Mental Mathematics.

Mental mathematics is the process of doing basic arithmetic off the top of your head, without the use of paper, and pencil, or an electronic device.

I will show you some techniques to do calculations, without paper and pencil or a calculator (a magic box), and if you practice enough, you may be able to do basic arithmetic, in some cases, even faster than a calculator.

About 35 years ago, when I attended the computer scientist conference in New Delhi, an Indian lady surprised the audience by doing basic math operations off the top of her head.

She could perform addition, subtraction, multiplication, and division of very big numbers even faster than calculators used during those days. Of course, I was amazed.

Though India is the land of magician and in everyday life on the road, you encounter crowds gathered around someone who is performing magic, to me, this one was a mystery, not magic.

Later, when I was director of Tehran Computer Service Center in Persia (present Iran), I met a student who appeared for university entrance exam competition, and he could perform fast basic math operation, similar to that Indian lady.

At this point, I decided to discover the mystery of their work, but it was not that easy, and there was no access to the internet yet. Therefore, I started my search in bookstores available in Tehran. After a long time, I found a math book in Persian that contained the answer to the mystery of mental fast calculation, or in other words, rapid arithmetic, originally written by Trachtenberg.

My book is basically a sort of modification of the work of Trachtenberg, a Russian engineer born in 1888 who was a political prisoner in Germany during Hitler's regime.

Here I have formulated the procedures of his work and used graphs and tables to show operations step by step. Technically, I have applied the algorithm, but I did not use Algol, or any other computer programming language to avoid difficulties for those who are not familiar with computer science and programming methodology.

I have included a few more techniques which are my own or have been collected from other sources. Unfortunately, I don't remember the names of any of them, but obviously many mathematicians from centuries ago were interested in this field, and it is even very much possible that the work of Trachtenberg is also a sort of modification of the past mathematicians' work.

Who are My Readers?

My readers are a variety of people with math backgrounds, those as young as an elementary school student or as old as an adult. If you are someone with a good mathematics background, may escape explanations and compliments that you already know.

Every language consists of letters and words with its own grammar; the case is the same with mathematics, science, and technology.

The mathematics language that we use in our daily lives, consists of decimal system digits (0- 9), numbers (combination of digits), some special characters such as +, and =, and much more.

Each operation in mathematics obeys certain rules, and the collection of these rules is the grammar of mathematics language.

The basic operations we do in mathematics are the addition, subtraction, multiplication, and division; even in the most simple math problems, we use at least one of these operations.

To perform any math operation correctly, we are bound to perform specific rules and regulations, referred as the grammar of mathematics language.

These rules are not written in stone; therefore whenever we discover a better new rule, we may use it, and gradually replace it with the one which is presently in use.

Trachtenberg's work may lead us to introduce some simple new rules to do basic math operations. These new rules are easy to perform, which helps to avoid mistakes much more than the present traditional rules used in school textbooks.

Table of Contents

Chapter 1 1

What is Multiplication? 1

    Multiplication Table 1

Multiplication of Zero and a Number 1

Multiplication of 1 and a Number 2

Multiplication of 2 and a Number 2

    The Most Significant Digit and the Least Significant Digit in a Number 3

Multiplication of 3 and a Number 5

Multiplication of 4 and a Number 7

Multiplication of 5 and a Multi-Digit Number 9

    A Special Case of Numbers Consisting of 9 Only 12

Multiplication of 6 and a Multi-Digit Number 13

    A Special Case of Numbers Consisting of 9 Only 14

Multiplication of 7 and a Multi-Digit Number 14

    A Special Case of Numbers Consisting of 9 Only 16

    Practice: 16

Multiplication of 8 and a Multi-Digit Number 17

Multiplication of 9 and a Multi-Digit Number 19

    Practice: 20

    Digit 9 Formula 20

Chapter 2 22

Two-Digit Numbers Multiplication 22

    •Special Case of Multiplication with Two-Digit Numbers: 25

    Multiplication of 11 and a Two-Digit Number 25

    •Multiplication of 25 and a number which is divisible by 4 26

    •Multiplication of 101 and a Two-Digit Number 27

    •Multiplication with Number 10 Families 28

    •Examining Multiplication Correctness 28

- ●Multiplication Correctness Rule 30
- ●Multiplying Two-Digit and Multi-Digit Numbers 30
- ●Multiplying Multi-Digit Numbers by Multi-Digit Numbers 32
- ●Multiplying Three-Digit Numbers by Three-Digit Numbers 33
- ●Multiple-Digit Numbers Multiplication- With No Modification 41
- ●Two Fingers Trick Procedure 41

Chapter 3 49
What is Addition? 49
- ●Examining Addition Correctness 54
- ●Add and Subtract Addition Method 56
- ●Correctness check of the Basic Math Operations 58
  - ●Digits Addition – Method 58
  - ●Number Eleven–Method 60
  - ●Two-Digit Numbers–Case 60
  - ●Multi-Digit Numbers–Case 60

Chapter 4 62
Division 62
- ●The Trachtenberg Division Method 62
- ●The Trachtenberg Division Procedure 62
- ●Correctness – Check 66
- ●Rapid Division Method 67
  - U + T Multiplication Method 68
  - UT + T Multiplication Method 68
- ●Division Operation with Three-Digit Divisor 75
  - Remainder Determination 82
Long Division 84

Chapter 5 90
The Square of a Number 90
The Square of a Two - Digit Number 90
Trachtenberg Rapid Method, Special numbers, Type 1 and Type 2 91

General Procedure of Squaring 2-Digit Numbers 92

Procedure of squaring Three-digit Numbers 93

The Square of an Arbitrary Three Digit Number 94

How to calculate the square of a typical number? 96

    Square of a number consisting of digit 1 only 96

    Square of numbers consisting of digit 2 only 97

    Square of numbers consisting of only digit 3, or 4, or...9 97

    Square and Multiplication Table of one Digit Numbers 98

Chapter 6 99

The Square Root of a Number 99

    Determining the three - digit and four - digit, numbers square - root 99

    Anticipating the Number of Digits in the Square Root of a Number 104

Chapter 7 114

Proof of Correctness 114

    Definition 7 – 1 114

    Definition 7 – 2 114

    Formula 7 – 1 115

    The Correctness Proof of Number N by 11 rules 115

Derive a multiplication formula for number N by 11, with m digits in N. 115

    The Correctness Proof of Number N by 6 rules 117

Derive a multiplication formula for number N by 6, with m digits in N.
Answer – 117

    The Correctness Proof of Number N by 5 rules 119

    The Correctness Proof of Number N by 9 rules 121

    The Correctness Proof of Number N by 8 rules 123

Multiplication Operation - Units Digit and Tens Digit Method 124

Multi Digit Numbers Multiplication – Units Digit and Tens Digit Method 126

Mental Mathematics

Chapter 1

In this chapter, we will introduce the Trachtenberg multiplication rules with the help of some examples.

But before introducing the Trachtenberg multiplication rules, we will become familiar with the multiplication table. This should be memorized so you can get rid of the calculator when doing basic multiplication such as 2×2 or 7×9, in other words, multiplication of a one-digit number by another one-digit number.

What is Multiplication?

It is basically a form of addition operation in which all numbers being used are really the same number added to one another repeatedly; for example, $23 \times 4 = 23 + 23 + 23 + 23 = 92$.

Multiplication Table

0	1	2	3	4	5	6	7	8	9
1	1	2	3	4	5	6	7	8	9
2	2	4	6	8	10	12	14	16	18
3	3	6	9	12	15	18	21	24	27
4	4	8	12	16	20	24	28	32	36
5	5	10	15	20	25	30	35	40	45
6	6	12	18	24	30	36	42	48	54
7	7	14	21	28	35	42	49	56	63
8	8	16	24	32	40	48	56	64	72
9	9	18	27	36	45	54	63	72	81

Once we memorize the above table, with a little bit of practice, we will be able to give the product of any two single-digit numbers without the use of a pen, pencil, or calculator. This is the foundation of our work in this book.

Multiplication of Zero and a Number

●The product of zero and a number, regardless of the number of digits, is always zero.

Example 1 – 1:

5 × 0 = 0.

19 × 0 = 0.

587,946 × 0 = 0.

Multiplication of 1 and a Number

●**Multiplying 1 by any number produces the same number.**

Example 1 – 2:

7 × 1 = 7

29 × 1 = 29

89,571,206 × 1 = 8,957,120

Multiplication of 2 and a Number

●**If every digit in our number is less than 5, as in 41,302, then to obtain its product when multiplied by 2, just add each digit by itself, or multiply each digit by 2.**

Example 1–3:

What is the product of 41,302 × 2?

With the help of the above rule, we can see:

2 + 2 = 4, or 2 × 2 = 4;

0 + 0 = 0, or 0 × 2 = 0;

3 + 3 = 6, or 3 × 2 = 6;

1 + 1 = 2, or 1 × 2 = 2;

4 + 4 = 8, or 4 × 2 = 8; and thus, the product of 41,302 × 2 = 82,604

Mental Mathematics

| 4 | 1 | 3 | 0 | 2 | × | 2 |

4 + 4 = 8	1 + 1 = 2	3 + 3 = 6	0 + 0 = 0	2 + 2 = 4
8	2	6	0	4

If our number contains digits equal to or more than 5, then the way we process changes. **Whenever a digit is 5 or more than 5, we get a 1 that must be carried over to the result of the next digit, adjacent to the left.**

The Most Significant Digit and the Least Significant Digit in a Number
- the least significant digit in a number is the digit located at its far right and the most significant digit is the digit located at its far left

For example, the least significant digit and the most significant digit of the number 25497658 respectively are 2, and 8.

Example 1 – 4:

What is the product of 713058 × 2?

Step 1

First, put a 0 to the left of 7, giving us: 0713 058 × 2

With the help of the above rule, we can continue:

Step 2

8 + 8 = 16 (6 is the least significant digit of our product and 1 is carried over to the product of the next digit to the left by 2).

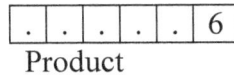
Product

Step 3

5 + 5 = 10 and then 10+1 (carryover) = 11.

Mental Mathematics

(The right-hand side 1 is the next digit of our product, and the left-hand 1 is carried over to the next step.)

.	.	.	.	1	6

Product

Step 4

0 + 0 = 0 and then 0 + 1 (1 is the carryover from Step 3) = 01.

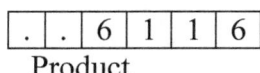

Step 5

3 + 3 = 6 and then 6 + 0 (0 is the carryover from Step 4) = 06.

.	.	6	1	1	6

Product

Step 6

1 + 1 = 2, and then 2 + 0 (0 is the carryover from Step 5) = 02.

.	2	6	1	1	6

Product

Step 7

7 + 7 = 14 and then 14 + 0 = 14 (1 is carried over to the next step)

4	2	6	1	1	6

Product

Step 8

0 + 0 = 0 and then 0 + 1 = 1 ⇨ **0713 058 × 2 = 1426116**

1	4	2	6	1	1	6

Mental Mathematics

Multiplication of 3 and a Number
General Rules:

Step 1

To obtain the least significant digit of the product, subtract the least significant digit from 10, then multiply it by 2; if it is an odd digit, add a 5 to it as well.

Step 2

Subtract every interior digit from 9 and then multiply it by 2. Then add the result to half of the right-adjacent digit, and if any digit is odd, add a 5 to its result also.

Step 3

To obtain the most significant digit of the product, divide the most significant digit by 2 and then subtract 2 from the result.

Note: If in any step, the result is a two-digit number, transfer the left digit to the next step.

Note: If every digit in a number is less than or equal to 3, such as in 231,023, then to obtain the product, just multiply each digit by 3.

Example 1 – 5:

What is 231023×3?

$3 \times 2 = 6, 3 \times 3 = 9, 3 \times 1 = 3, 3 \times 0 = 0, 3 \times 2 = 6$, and $3 \times 3 = 9$,

⇨ **231023 × 3 = 693069**

6	9	3	0	6	9

Product

Example 1 – 6:

What is the product of 2478×3?

Mental Mathematics

Step 1

First, put a 0 to the left of 2, giving us: 02478 × 3?

Step 2

10 − 8 = 2 and then 2 × 2 = 4.

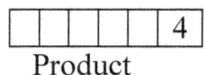
Product

Step 3

9 − 7 = 2 and then 2 × 2 = 4 and then 4 + 8 / 2 = 4 + 4 = 8 and then 8 + 5 = 13.

(We added 5 to the result because 7 is an odd digit).

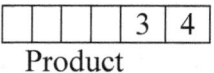
Product

Step 4

9 − 4 = 5 and then 2 × 5 = 10 and then 10 + 7 / 2 = 10 + 3 = 13, and then 13 + 1(carryover from Step 3) = 14.

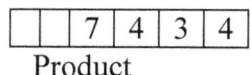
Product

Step 5

9 − 2 = 7 and then 2 × 7 = 14 and then 14 + 4 / 2 = 14 + 2 = 16 and then 16 + 1 (carryover from Step 4) = 17.

| | | 7 | 4 | 3 | 4 |
Product

Step 6

2 / 2 = 1 and then 1 + 1(carryover from Step 5) = 2 and ⇨ 2 − 2 = 0.

| Product = | 0 | 7 | 4 | 3 | 4 |

Mental Mathematics

Multiplication of 4 and a Number
General Rule:

Step 1

Put a zero on the left-hand side of your multi-digit Number.

Step 2

Subtract the least significant digit from 10 to obtain the least significant digit of the product. If any digit in our number is an odd digit; we have to add a 5 to it as well.

Step 3

Subtract every interior digit from 9, and add the result to the half of the right-hand side digit.

Step 4

When we reach the zero on the left-hand side of the most significant digit of our number, we do an operation on the most significant digit:

To obtain the most significant digit of a product, subtract 1 from half of the most significant digit.

Example 1–7:

What is the product of 7439805 × 4?

Step 1

First, put a 0 on the left of 7, giving us: 07439805 × 4?

Step 2

10 – 5 = 5 and then 5 + 5 = 10 we added a 5 to the result of 10 – 5, because 5 is an odd digit (1 is carryover to Step 3 and zero is the least significant digit of the product).

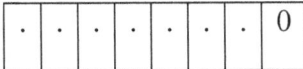

Mental Mathematics

Step 3

Subtract every interior digit from 9 and add the result to the half of the right-hand side digit.

9 − 0 = 9 and then 9 + 5 / 2 = 9 + 2 = 11 + 1 (carryover from Step 2) = 12.

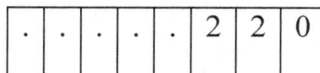
Product

Step 4

9 − 8 = 1 and then 1 + 0 / 2 = 1 + 0, and 1 + 1 (carryover from Step 3) = 2.

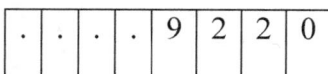
Product

Step 5

9 − 9 = 0 and then 0 + 8 / 2 = 0 + 4 = 4 and then 4 + 5 = 9.

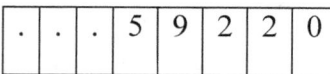
Product

Step 6

9 − 3 = 6 and then 6 + 9 / 2 = 6 + 4 = 10 and then 10 + 5 = 15(1 of 15 is carried over to Step 7).

.	.	.	5	9	2	2	0

Product

Step 7

9 − 4 = 5 and then 5 + 3 / 2 = 5 + 1 = 6 and then 6 + 1 (carryover from Step 6) = 7.

.	.	7	5	9	2	2	0

Product

Mental Mathematics

Step 8

9 − 7 = 2 and then 2 + 4 / 2 = 2 + 2 = 4 and then 4 + 5 = 9.

.	9	7	5	9	2	2	0

Product

Step 9

When we reach the zero on the left-hand side of the most significant digit of our number, we do an operation on the most significant digit:

To obtain the most significant digit of the product, subtract 1 from half of the most significant digit.

7 / 2 − 1 = 3 − 1 = 2 ⇨ 7439805×4 =

2	9	7	5	9	2	2	0

Multiplication of 5 and a Multi-Digit Number

Part A: If all digits of our number are even (0, 2, 4, 6, and 8), its product is obtained by dividing each digit by 2, and putting a zero on the right of the least significant digit.

Example 1 − 8:

What is the product of 6482208 × 5?

Start from left and proceed to right 6 / 2 =3, 4 / 2 = 2, 8 / 2 = 4, 2 / 2 = 1, 2 / 2 = 1, 0 / 2 = 0, 8 / 2 = 4, therefore: 6482208 × 5 = 32411040.

Part B: If all digits in a number are odd (1, 3, 5, 7, or 9), to obtain the product of such a number by 5, we start from the right digit and divide its right adjacent digit by 2, add a 5 to it, and proceed to left, doing the same with all digits.

Mental Mathematics

Example 1 – 9:

What is the product of 35975 × 5?

Step 1

First, put a 0 to the left of 3, giving us: 035975 × 5?

Step 2

The least significant digit of 035975 is 5 and there is no digit on the right side of 5, so we assume it to be zero, therefore least significant digit of our product is 0 / 2 + 5 = 5.

Step 3

7 is the next digit and its adjacent right digit is 5, therefore the next digit of our product is 5/ 2 + 5 = 2 + 5 = 7.

Note: We always assume 1 / 2 = 0, 3/2 = 1, 5 / 2 = 2, 7 / 2 = 3 and 9 / 2 = 4.

Step 4

9 is the next digit and its adjacent right digit is 7, therefore the next digit of our product is 7 / 2 + 5 = 3 + 5 = 8.

Step 5

5 is the next digit and its adjacent right digit is 9, therefore the next digit of our product is 9 / 2 + 5 = 4 + 5 = 9.

Step 6

3 is the next digit and its adjacent right digit is 5, therefore the next digit of our product is 5 / 2 + 5 = 2 + 5 = 7.

Step 7

0 is the next digit and its adjacent right digit is 3, therefore the next digit of our product is 3 / 2 (We do not add 5 because 0 is considered to be an even digit) = 1.

Mental Mathematics

> ⇨ Product = 179875.

Part C: If a number consists of even and odd digits, we use a rule which is a combination of rules in part A and part B.

Example 1–10:

What is the product of 65478892 × 5?

Step 1

First, put a 0 to the left of 6, giving us: 065478892 × 2.

Step 2

Least significant digit of our number is 2, and there is no adjacent digit on its right-hand side, so we assume it to be 0, and therefore the least significant digit of our product is 0 / 2 = 0.

Step 3

The next digit to be considered is 9 and 2 is the next right adjacent digit to it, which is an odd digit, therefore 2 / 2 + 5 = 6 is the next digit of our product.

Step 4

The next digit to be considered is 8 and 9 is the next right adjacent digit to it, which is an even digit, therefore 9 / 2 = 4 is the next digit of our product.

Step 5

The next digit to be considered is 8 and 8 is also the next right adjacent digit to it, which is an even digit, therefore 8 / 2 = 4 is the next digit of our product.

Step 6

The next digit to be considered is 7 and 8 is the next right adjacent digit to it, which is an even digit, therefore 8 / 2 + 5 = 9 is the next digit of our product.

Step 7

The next digit to be considered is 4 and 7 is the next right adjacent digit to it, which is an even digit, therefore $7 / 2 = 3$ is the next digit of our product.

Step 8

The next digit to be considered is 5 and 4 is the next right adjacent digit to it, which is an even digit, therefore $4 / 2 + 5 = 7$ is the next digit of our product.

Step 9

The next digit to be considered is 6 and 5 is the next right adjacent digit to it, which is an even digit, therefore $5 / 2 = 2$ is the next digit of our product.

Step 10

The last digit to be considered is 0 (an even digit), and 6 is the next right adjacent digit to it, therefore $6 / 2 = 3$ is the next digit of our product, and thus,

$$065478892 \times 5 = 327394460.$$

Note: There are always special cases that make our job much easier, and in no time, we may be able to find the product of a digit and a multi-digit number.

In many cases, I will show you some example of special cases, but I left a good number of them for you as exercises to improve your math ability.

A Special Case of Numbers Consisting of 9 Only

Now let's consider the product of a multi-digit number which consists of 9 only by 5.

What is the product of 999999×5?

Whenever multi–digit numbers consist of only 9, change the least significant digit to 5, and put a 4 on the left-hand side of the most significant digit to obtain the product, thus **$999999 \times 5 = 4999995$**.

Obviously, if you use the general rules explained earlier, you will get the same result.

Mental Mathematics

Multiplication of 6 and a Multi-Digit Number
General Rule:

Begin from right to left, and add each digit to the half of the right adjacent digit, and add a 5 to any digit which is odd.

Example 1 – 11:

What is the product of 349 × 6?

Step 1

Put a 0 to the left of 3, giving us: 0349 × 6.

Step 2

9 + 5 + 0 / 2 = 14 (4 is the least significant number of product, and 1 is the carryover digit to the next step)

Step 3

4 + 9 / 2 + 1 (1 is the carryover from Step 1) = 4 + 4 + 1 = 9.

Step 4

3 + 5 + 4 / 2 = 10

Step 5

0 + 3 / 2 + 1 = 2 ⇨ 349×6 = 294

$\boxed{349×6 = 2094}$

Example 1 – 12:

What is the product of 85923 × 6?

Step 1

Put a 0 to the left of 8, giving us: 085923

Step 2

3 + 5 + 0/2 = 8

Step 3

2 + 3 / 2 = 2 + 1 = 3

Step 4

9 + 5 + 2 / 2 = 15

Step 5

5 + 5 + 9 / 2 + 1 (carryover from Step 4) = 10 + 4 + 1 = 15.

Step 6

8 + 5 / 2 + 1 (carryover from Step 5) = 8 + 2 + 1 = 11.

Step 7

0 + 8 / 2 + 1 (carryover from Step 6) = 4 + 1 = 5 ⇨ **85923 × 6 = 515538.**

Product = 515538

A Special Case of Numbers Consisting of 9 Only
Whenever a multi-digit number consists of digit 9 only, change the least significant digit to 4 and put a 5 on the left-hand side of the most significant digit to obtain the product ⇨ 99999999 × 6 = 599999994.

Multiplication of 7 and a Multi-Digit Number
General Rule:

Begin from right to left and multiply each digit by 2, and then add half of the right adjacent digit and add a 5 to any digit which is odd.

Example 1 – 13:

What is the product of 4242 × 7?

Mental Mathematics

Step 1

First, put a 0 to the left of 4, giving us: 04242 × 7?

Step 2

2 × 2 = 4 ⇨ 4 + 0 / 2 = 4

Step 3

2 × 4 + 2 / 2 = 9

Step 4

2 × 2 + 4 / 2 = 6

Step 5

2 × 4 + 2 / 2 = 9

Step 6

2 × (0) + 4 / 2 = 2

In this example all digits are even, therefore no 5 is added.

Product = 29694

Example 1 – 14:

What is the product of 85432 × 7?

Step 1

First, put a 0 to the left of 8, giving us: 085432 × 7?

Step 2

2 × 2 + 0 / 2 = 4

Mental Mathematics

Step 3

2 × 3 + 2 / 2 + 5 = 12 (1 is carryover to next Step)

Step 4

2 × 4 + 3 / 2 + (1 is carryover from Step 3) = 8 +1 +1 =10.

Note: Technically 3 / 2 = 1.5, but it is rounded to 1 and we use the same rule for all odd digits.

Step 5

2 × 5 + 5 + 4 / 2 + 1 (1 is carryover from Step 4) = 18.

Step 5:

2 × 8 + 5 / 2 + 1 (1 is carryover from Step 5) = 16 +2 + 1 = 19.

Step 6:

2(0) + 8 / 2 + 1 (1 is carryover from Step 5) = 5.

Product = 598024.

A Special Case of Numbers Consisting of 9 Only
Whenever a multi-digit numbers consist of only 9, change the least significant digit to 3 and put a 6 on the left-hand side of the most significant digit to obtain the product.
⇨ 99999999 × 7 = 699999993

Practice:
What is the product of each the following?

1) 278593344105 × 5 =?
2) 846202486002 × 5 =?
3) 999753977553 × 5 =?
4) 537986400884 × 6 =?
5) 846202486002 × 6 =?
6) 789753977553 × 7 =?
7) 578595544105 × 7 =?
8) 789583977553 × 6 =?

Mental Mathematics

9) 346202486002 × 6 =?

Multiplication of 8 and a Multi-Digit Number

Example 1 – 15:

What is the product of 867 × 8?

Step 1

First, put a 0 to the left of 8, giving us: 0867 × 8?

Step 2

Subtract the least significant digit from 10, and then multiply the result by 2.

⇨ 10 − 7 = 3 and then 3 × 2 = 6, and 6 is the least significant digit of our product.

Step 3

Subtract interior digits from 9 and then multiply by 2. Then add it to the right adjacent digit, and if the result is a two-digit number, then the right digit is the second digit of the product, and the left digit is the carryover to the next step.

⇨ 9 − 6 = 3, and then 2 × 3 = 6, and then 6 + 7 = 13 (1 is carryover to the next step), and 3 is the next digit of our product at the left of 6.

Step 4

Since there is a zero to the left of 8, this digit is also considered as an interior digit, so we do the same to 8 as we did with 9 in Step 3.

So we write 9 − 8 = 1, and then 2 × 1 = 2 and then 2 + 6 = 8, adding carryover 1 to it, 8 + 1 = 9 and 9 is the next digit of our product.

Step 5

Subtract 2 from the most significant digit and add it to carryover from the previous step.

So we write 8 − 2 = 6, and carryover from Step 4 is zero.

Mental Mathematics

> Product of 867 × 8 = 6936.

Example 1 – 16:

What is the product of 987352 × 8?

Step 1

First, put a 0 to the left of 9, giving us: 0987352 × 8?

Step 2

Zero and 2 are exterior digits and the other digits are interior digits.

Step 3

10 – 2 = 8, and then 8 × 2 = 16 (6 is the least significant digit of our product and 1 is carried over to Step 3)

Step 3

9 – 5 = 4, and then 2 × 4 = 8, and then 8 + 2 + 1(carryover from Step 2) = 11 in which right 1 is the next digit of our product at the left of 6, and left 1 is the carryover.

Step 4

9 – 3 = 6, and then 2 × 6 = 12, and then 12 + 5 +1 (carryover from Step 3) = 18, in which 8 is the next digit of our product at the left of 1, and 1 is the carryover.

Step 5

9 – 7 = 2, and then 2 × 2 = 4, and then 4 + 3 + 1 (carryover from Step 4) = 8, in which 8 is the next digit of our product at the left of 8, and 0 is the carryover.

Step 6

9 – 8 = 1, and then 2 × 1 = 2, and then 2 + 7 = 9, in which 9 is the next digit of our product at the left of 8, and 0 is the carryover.

Mental Mathematics

Step 7

9 − 9 = 0, and then 2 × 0 = 0 and then 0 + 8 = 8, in which 8 is the next digit of our product at the left of 9, and 0 is the carryover.

Step 8

Subtract 2 from the most significant digit and add it to the carryover from the previous step. 9 − 2 = 7, and then 7 + 0 = 7, therefore 7 is the most significant digit of our product.

⇨ Product = 987352 × 8 = 7898816

Multiplication of 9 and a Multi-Digit Number

Example 1 – 17:

What is the product of 7564 × 9?

Step 1

First, put a 0 to the left of 7, giving us: 07564 × 9?

Step 2

Subtract the least significant digit from 10 to obtain the least significant digit of the product.

10 − 4 = 6.

Product = ………..6

Step 3

To obtain the interior digit of our product, we subtract each interior digit from 9 and then add it to the right adjacent digit.

9 − 6 = 3, and then 3 + 4 = 7.

Product = …………76

Mental Mathematics

Step 4

9 − 5 = 4, and then 4 + 6 = 10 (1 is carryover to Step 5)

Product = …………..076

Step 5

9 − 7 = 2, and then 2 + 5 + 1 (carryover from Step 4) = 8.

Step 6

Subtract 1 from the most significant digit, and add the carryover to it.

7 − 1 = 6 and then 6 + 0 = 6.

> Product of 07564 × 9 = 68076

Practice:
What is the product of each the following?

1) What is 871695 × 8 =?
2) What is 397441 × 9 =?
3) What is 947221 × 7 =?
4) What is 15783 × 6 =?
5) What is 7442319 × 5 =?
6) What is 8466240 × 5 =?
7) What is 15783 × 9 =?
8) What is 947221 × 8 =?
9) What is 845221 × 3 =?
10) What is 45783 × 4 =?
11) Choose a number and multiply it to 0, 1, 2, 3, 4, 5, 6, 7, 8, and, 9.

Digit 9 Formula
In general, to obtain the product of a number consisting of M digit 9 by a single digit number, we may use the following formula:

99999…×N = [N − 1][$9_{(M-1)}9_{(M-2)}9_{(M-3)}…9_{(M-(M)-1)}$][10 − N]

Mental Mathematics

Example 1 – 18

What is 99999×8?

In this example N = 8 and M = 5 ⇨ 99999×8 = [8 – 1][9999][10 – 8] = 799992.

We may write the above formula as below:

$[9_M 9_{(M-1)} 9_{(M-2)} \ldots 9_{(M-(M-1))}] \times N = [N-1][9_{(M-1)} 9_{(M-2)} 9_{(M-3)} \ldots 9_{(M-(M)-1)}][10-N]$.

Mental Mathematics

Chapter 2

In this chapter, we begin our work with the multiplication of two-digit numbers and become familiar with some primary multiplication technique.

Product = (Multiplier) × (Multiplicand)

Let's take an example and explain the terminology we need to use here:

Two-Digit Numbers Multiplication

Example 2 – 1:

What is the product of 28 × 37?

To find the above product with the help of Trachtenberg Rules, we do the following:

Step 1

Multiply the least significant digits of multiplicand and multiplier, as below:

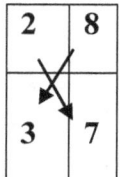

8 × 7 = 56, where 6 is the last digit of the product and 5 is the carryover.

Step 2

Now cross multiply and add them as follows:

2	8
3	7

Mental Mathematics

$2 \times 7 + 8 \times 3 = 14 + 24 = 38$, and then add the carryover from previous step:

⇨$38 + 5 = 43$⇨3 is the next digit of our product, and 4 is the carryover.

Step 4

Multiply the most significant digit of multiplicand and multiplier, and add 4 to it, which is the carryover from Step 2:

2	8
3	7

$2 \times 3 + 4 = 10$.

0 is the next number of the product, and 1 is carried to the left, ⇨ $1 + 0 = 1$, and 1 is the most significant digit of the product.

Product =	1	0	3	6

Example 2 – 2:

What is the product of 47×52?

Step 1

$47 \times 52 = 0047 \times 52$.

Multiply the least significant digits of multiplicand and multiplier

4	7
5	2

Mental Mathematics

$7 \times 2 = 14$, thus 4 is the least significant digit of the product, and 1 is carried to next column.

Step 2

Now cross multiply and add them as follows:

4	7
5	2

$4 \times 2 + 7 \times 5 + 1$ (carryover from Step 1) = 44, thus 4 (the least significant digit of 44) is the next digit of product, and 4(the most significant digit of 44) is the carryover.

Step 3

Multiply the most significant digit of multiplicand and multiplier, and add 4 to it, which is the carryover digit from Step 2:

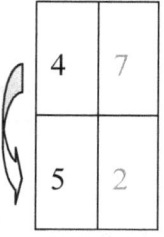

4	7
5	2

$4 \times 5 + 4 = 24$, thus 4 is the next digit of product, and 2 the carryover to the next column. Added to zero, most significant digit of the product is 2.

| Product = | 2 | 4 | 4 | 4 |

Mental Mathematics

•Special Case of Multiplication with Two-Digit Numbers:

Multiplication of 11 and a Two-Digit Number
To figure out the product of 11 and a two-digit number (MN), we will use a couple of simple formulas. For the two different cases, we assume M + N = Z.
1) **If Z ≤ 9 ⇨ Product = MZN, where Z = (M + N).**

Example 2 – 3:

What is the product of 11 × 27?

M = 2 and N = 7 ⇨ Z = 2 + 7 = 9,

⇨ **Product = MZN = 0297.**

Example 2 – 4:

What is the product of 11 × 34?

M = 3, N = 4 and thus Z = 3 + 4 = 7 ⇨ **Product = MZN = 374.**

2) If Z > 9, then product = BFN, where B = (M + C) and Z = CF.

Example 2 – 5:

What is the product of 11×39?

M = 3, N = 9, and thus Z = CF = 3 + 9 = 12, and thus, C = 1, and F = 2.

B = (M + C) = 3 + 1 = 4 ⇨ **Product = BFN = 0429.**

Example 2 – 6:

What is the product of 11 × 99?

M = 9, N = 9 and thus Z = CF = 9 + 9 = 18, and thus, C = 1, F = 8 and B = (M + C) = 9 + 1 = 10 ⇨ **Product = BFN = 1089.**

Mental Mathematics

Example 2 – 7:

What is the product of 11 × 68?

M = 6, N = 8, and thus Z = CF = 6 + 8 = 14, and thus C = 1, F = 4 and B = (M + C) = 6 + 1 = 7 ⇨ **Product = BFN = 00748.**

●**Multiplication of 25 and a number which is divisible by 4**

In general, if we multiply 25 into a number which is divisible by 4, its product is obtained by dividing it by 4 and multiplying it by 100 (insert two zeros on the right of its last digit).

What is the product of 4444 × 25?

Though we may use the formula introduced for two-digit and a multi-digit number, there is a shortcut to find the product of 25 and numbers with multiples of 4:

Change each 4 to a 1 and insert two zeros in front of the last digit.

44 × 25 = 1100, 444 × 25 = 11100, 4444 × 25 = 111100, and 44444444 × 25 = 1111111100 and so on and so forth

Example 2 – 8:

See the product of the following multiplications:

1) 324 × 25

⇨324 / 4 = 81 ⇨product = 8100.

2) 640× 25

⇨640 / 4 = 160 ⇨product = 16000.

3) 16 × 25 ⇨16 / 4 = 4 ⇨product = 400.

Mental Mathematics

•Multiplication of 101 and a Two-Digit Number

To obtain the product of 101 and an arbitrary two-digit number, simply put the same number in front of your arbitrary two digit number. For example, if MN is your arbitrary two digit number then 101 × MN = MNMN (where M and N are single digit numbers).

Example 2 - 9:

See the product of the following multiplications:

1) 101×67 = 6767

2) 101×99 = 9999.

3) 101×11 = 1111.

4) 101×53 = 5353

We may use the same method to find the product of 1001 × MNK and 10001 × MNKL, (where M, N, K, and L are single digit numbers):

1001 × MNK = MNKMNK.

10001 × MNKL = MNKLMNKL

Example 2 – 10:

See the product of the following multiplications:

1) 1001×532 = 532532.

2) 1001×985 = 985985.

3) 10001×9953 = 99539953.

4) 10001×4978 = 49784978.

Mental Mathematics

You may use the same method to find the product of bigger numbers with the same nature:

10000001 × 9587629 = 95876299587629

• **Multiplication with Number 10 Families**

To obtain the product of an arbitrary number by 10, 100, 1000, and so on, simply put as many zeros as you have in the 10 families in front of your arbitrary number.

Example 2 – 11:

10 × 258 = 2580.

100 × 981 = 98100.

1000 × 9212 = 9212000.

• **Examining Multiplication Correctness**

The technique of examining multiplication correctness described here is not introduced by Trachtenberg, it is an old technique used by mathematicians for many centuries. Once you have an understanding of it, it is very easy to use.

As I explained before in multiplication operation, we encounter with three factors: multiplicand, multiplier, and product.

Each of these factors could be a single digit number or a multi-digit number.

In 2 ×3 = 6, 3 × 3 = 9, and 4 × 2 = 8, all multiplicand, multiplier, and product are single digit numbers, but in 9 × 5 =45, multiplicand and multiplier are single digit numbers and the product is a two-digit number, while in 569 × 387 = 220203, all are multi-digit numbers.

Mental Mathematics

To explain the technique of multiplication correctness, the example will have multi-digit numbers for the multiplicand, multiplier, and product.

Multiplicand = 8954736.
Multiplier = 78954.

Product = 8954736 × 78954 = 707012226144.

The following steps show how this product was found:

Step 1

Add up all digits in multiplicand to reach to a single digit number:

8 + 9 = 17, and 1 + 7 = 8, and 8 + 5 = 13, and, 1 + 3 = 4, 4 + 4 = 8, and 8 + 7 = 15, and 1 + 5 = 6, and 6 + 3 = 9, and 9 + 6 = 15, and 1 + 5 = 6

Note: If you add up the all digits, your final answer is the same:

8 + 9 + 5 + 4 + 7 + 3 + 6 = 42, and 2 + 4 = 6

Step 2

Add up all digits in multiplier to reach to a single digit number:

7 + 8 = 15, and 1 + 5 = 6, and 6 + 9 = 15, and 1 + 5 = 6, and 6 + 5 = 11, and 1 + 1 = 2, and 2 + 4 = 6

Step 3

Add up all digits in Product to reach to a single digit number:

7 + 0 = 7, 7 + 7 = 14, and 1 + 4 = 5, and 5 + 0 = 5, and 5 + 1 = 6, and 6 + 2 = 8, and 8 + 2 = 10, and 1 + 0 = 1, and 1 + 2 = 3, and 3 + 6 = 9, and 9 + 1 = 10, add 1 + 0 = 1, and 1 + 4 = 5, and 5 + 4 = 9.

Mental Mathematics

Step 4

Now multiply the single-digit obtained from multiplicand by the single-digit obtained from the multiplier ⇨ 6×6 = 36, and 3 + 6 = 9.

Since this is the same number you obtained by adding the digits of your product, your multiplication answer is correct.

• **Multiplication Correctness Rule**

Add up multiplicand digit. If the result of the addition is not a single digit, repeat the same thing with your new number until you get a single digit. Do the same with the multiplier, and then multiply the single-digit result from multiplicand by the single digit from the multiplier. If the result is not a single digit, add up its digits to reach a single-digit, and call it M. Then add the product digits. If the result is not a single digit, add up digits of your product result until you reach to a single digit, and call it N.

If M = N, then the result of your multiplication is correct.

• **Multiplying Two-Digit and Multi-Digit Numbers**

To multiply a two-digit number and a multi-digit number, we assume MN as our two-digit number, and DEFGH as a multi-digit number, where each of the M, N, D, E, F, G, and H is assumed to be a single-digit number, and C represents our carry number. (DEFGH and MN are the multiplicand and multiplier).

Step 1

a) 00DEFGH × MN = ILJKPRS (ILJKPRS is assumed to be the product of the multiplicand and multiplier).

b) If N×H = C_1S, then S is the least significant digit of product, and C_1 is the carryover to the next column ($C_1 = 0$ or $C_1 \neq 0$).

Step 2

Pair MN with every two digits of multiplicand: GH, FG, EF, DE, 0D, and 00.

Mental Mathematics

Perform the following operations:

a) For MN and GH write: $N \times G + M \times H + C_1 = C_2R$, where R is the adjacent digit to S, and C_2 is the carryover to the next column.
b) For MN and FG write: $N \times F + M \times G + C_2 = C_3P$, where P is the adjacent digit to R, and C_3 is the carryover to the next column.
c) For MN and EF write: $N \times E + M \times F + C_3 = C_4K$, where K is the adjacent digit to P, and C_4 is the carryover to the next column.
d) For MN and DE write: $N \times D + M \times E + C_4 = C_5J$, where J is the adjacent digit to K, and C_5 is the carryover to the next column.
e) For MN and 0D write: $N \times (0) + M \times D + C_5 = C_6L$, where L is the adjacent digit to J, and C_6 is the carryover to the next column.
f) For MN and 00 write: $N(0) + M(0) + C_6 = C_7I$, where I is the adjacent digit to L, and C_7 is the carryover to the next column ($I = C_6$).

For this problem, we are done here, but if the length of the multiplicand is longer, we need to continue on in the same way.

Example 2 – 12:

What is the product of 7345 × 23?

Step 1

Count number of digits in the multiplier and put the same number of zeros on the left-hand side of multiplicand.

007345 × 23 =?

Step 2

Multiply the least significant multiplier digit by least significant multiplicand digit:

3 × 5 = 15, and 5 is the least significant digit of our product, and 1 is the carryover to next step.

Step 3

Pair 23 with 45 and multiply them: 2 × 5 + 3 × 4 + 1 (carryover from Step 2) = 10 +12 + 1 = 23, where 3 is the adjacent digit to 5, and 2 is carryover.

Mental Mathematics

Step 4

Pair 23 with 34 and multiply them: $2 \times 4 + 3 \times 3 + 2$ (carryover from Step 3) = 8 + 9 + 2 = 19, where 9 is the adjacent digit to 3 and 1 is carryover.

Step 5

Pair 23 with 73 and multiply them: $2 \times 3 + 3 \times 7 + 1$ (carryover from Step 4) = 6 + 21 + 1 = 28, where 8 is the adjacent digit to 9 and 2 is the carryover.

Step 6

Pair 23 with 07 and multiply them: $2 \times 7 + 3 \times 0 + 2$ (carryover from Step 5) = 14 + 0 + 2 = 16, where 6 is the adjacent digit to 8 and 1 is the carryover.

Step 7

Pair 23 with 00 and multiply them: $2 \times 0 + 3 \times 0 + 1$ (carryover from Step 5) = 0 + 0 + 1 = 1.

1 is the most significant digit of the product.

Final Product = 168935.

•Multiplying Multi-Digit Numbers by Multi-Digit Numbers

We recognize a multi-digit number as any number with more than two digits, such as 529, 74100685, 5879201, 978412035 and so on. 452 × 873, 8972103 × 19548, and 78465 × 5807 are examples of multi-digit multiplication.

To avoid confusion, we always assume the multiplicand to be greater than the multiplier.

The procedure explained here is a modification of Trachtenberg rapid mental arithmetic trick, but with the help of algorithm and technology, it has become more organized and easier to understand. Once you understand the procedure, you do not have to do everything I do here, but you do need to practice as much as you can because the practice is the key to your success.

Mental Mathematics

Here I have defined two sets of digits which are multiplied and added according to certain rules and regulations to obtain the product of our multiplication operation:

1) Set of Multiplier – This is simply a set of digits in our multiplier. For example, if we are multiplying 8972103 × 19548, then 19548 is the set of the multiplier with 5 elements.

2) Set of Pointer – This is a set with the same number of elements as the set of the multiplier, which is represented by English alphabet letters, with each letter of the pointer pointing to a digit in the multiplicand. As we start our multiplication operation, the set of pointer keeps moving from the last digit of the multiplicand toward its first digit (moving from right to left).

● **Multiplying Three-Digit Numbers by Three-Digit Numbers**

Example 2 – 13:

What is the product of 583 × 234?

Step 1

Multiplicand = 583 and Multiplier = 234.

We put 3 zero on the left of the multiplicand (the number of zeros to the left of the multiplicand is always equal to the number of digits in multiplier).

			A	B	C			
0	0	0	5	8	3		2	3	4
	P	R	O	D	U	C	T	
								2

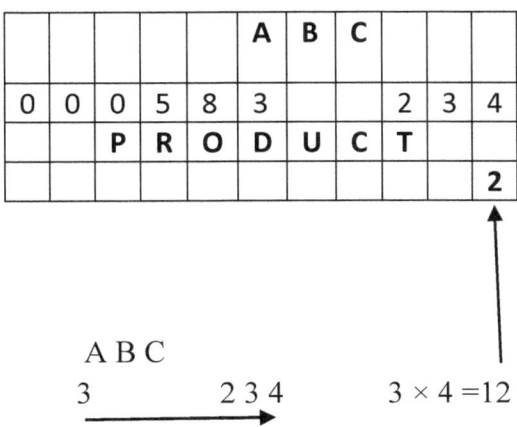

A B C
3   2 3 4   3 × 4 = 12

Digit 2 is the least significant of the product, and 1 is carried over to the next step.

Mental Mathematics

Step 2

			A	B	C			
0	0	0	5	8	3		2	3	4
		P	R	O	D	U	C	T
							2	2

8 3 2 3 4

8 × 4 + 3 × 3 + 1 (carryover from previous step) = 32 + 9 + 1 = 42, (2 is the next digit of the product, with 4 as the carryover to the next step).

Step 3

			A	B	C			
0	0	0	5	8	3		2	3	4
		P	R	O	D	U	C	T
						4	2	2

$$\underset{5\ 8\ 3}{\overset{A\ B\ C}{\longleftrightarrow}} \longleftrightarrow 2\ 3\ 4$$

3 × 2 + 8 × 3 + 5 × 4 + 4 (carryover from previous step) = 6 + 24 + 20 + 4 = 54, (4 is the next digit of the product, with 5 as a carryover to the next step).

Step 4

			A	B	C			
0	0	0	5	8	3		2	3	4
		P	R	O	D	U	C	T
					6	4	2	2

Mental Mathematics

```
    A B C
    0 5 8  ←——————→  2 3 4
```

$8 \times 2 + 5 \times 3 + 0 \times 4 + 5$ (carryover from previous step) $= 16 + 15 + 0 + 5 = 36$,

(6 is the next digit of the product, with 3 as the carryover to the next step.)

Step 5

	A	B	C						
0	0	0	5	8	3		2	3	4
		P	R	O	D	U	C	T	
					3	6	4	2	2

```
    0 0 5  ←——————→  2 3 4
```

$5 \times 2 + 0 \times 3 + 0 \times 4 + 3$ (carryover from Step 4) $= 10 + 0 + 0 + 3 = 13$, (3 is the next digit of the product, with 1 as the carryover to the next step).

Step 6

A	B	C							
0	0	0	5	8	3		2	3	4
		P	R	O	D	U	C	T	
			1	3	6	4	2	2	

```
  A B C
  0 0 0  ←——————→  2 3 4
```

$0 \times 2 + 0 \times 3 + 0 \times 4 + 1$ (carryover from Step 5) $= 0 + 0 + 0 + 1 = 1$,

(1 is the most significant digit of the product).

Mental Mathematics

Product = 136422.

Example 2 – 14:

What is the product of 75902481 × 9723?

Step 1

ABCD is the pointer set, which keeps moving one square toward the left in every step, and as we proceed, multiplicand digits under ABCD are paired with multiplier digits, as shown in each step.

									A	B	C	D				
0	0	0	0	7	5	9	0	2	4	8	1			×	9	7	2	3
P	R	O	D	U	C	T											3

A B C D   Multiplier Set = 9723

1   9723 → 1 × 3 = 3 (3 is the least significant digit of the product, with 0 as the carryover to the next step)

Step 2

								A	B	C	D					
0	0	0	0	7	5	9	0	2	4	8	1			×	9	7	2	3
P	R	O	D	U	C	T										6	3

ABCD (9723 is the multiplier set, which is paired with multiplicand digits under pointer ABCD).

8 1   9 7 2 3 → 1 × 2 + 8 × 3 = 2 + 24 = 26

(6 is the next digit of the product, with 2 as the carryover to the next step).

Mental Mathematics

Step 3

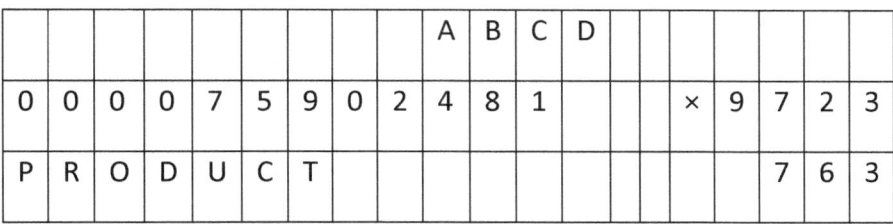

							A	B	C	D					
0	0	0	0	7	5	9	0	2	4	8	1		×	9	7	2	3
P	R	O	D	U	C	T								7	6	3

ABCD　　　　　　　　Multiplier Set = 9723

4 8 1　　　　　　　　9 7 2 3 → 4 × 3 + 8 × 2 + 1 × 7 + 2 (carryover from Step 2)

= 12 + 16 + 7 + 2 = 37. (7 is the next digit of the product, with 3 as the carryover to the next step).

Step 4

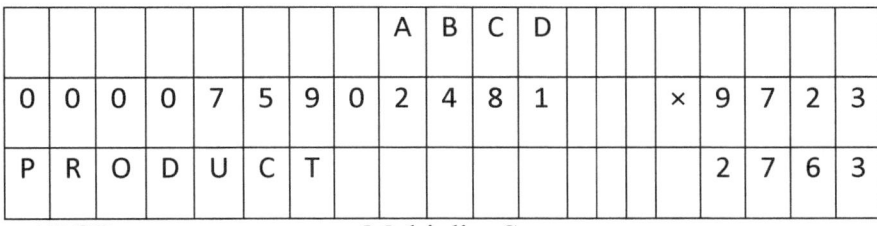

							A	B	C	D					
0	0	0	0	7	5	9	0	2	4	8	1		×	9	7	2	3
P	R	O	D	U	C	T								2	7	6	3

ABCD　　　　　　　　Multiplier Set

2 4 8 1 ────────→ 9 7 2 3 → 2 × 3 + 4 × 2 + 8 × 7 + 1 × 9 + 3

(3 is the carryover from Step 3) = 6 + 8 + 56 + 9 + 3 = 82. (2 is the next digit of the product, with 8 as the carryover to the next step).

Step 5

							A	B	C	D					
0	0	0	0	7	5	9	0	2	4	8	1		×	9	7	2	3
P	R	O	D	U	C	T							2	2	7	6	3

Mental Mathematics

A B C D   Multiplier Set = 9723

0 2 4 8 ⟶ 9 7 2 3 ⇒ 0×3 + 2×2 + 4×7 + 8×9 + 8 (Carryover from step 4) = 0 + 4 + 28 + 72 + 8 = 112. (2 is the next digit of the product, with 11 as the carryover to the next step).

Step 6

						A	B	C	D						
0	0	0	0	7	5	9	0	2	4	8	1		×	9	7	2	3
P	R	O	D	U	C	T						8	2	2	7	6	3

ABCD   Multiplier Set = 9723

9 0 2 4 ⟶ 9 7 2 3 → 9 × 3 + 0 × 2 + 2 × 7 + 4 × 9 + 11 (carryover from

Step 5) = 27 + 0 + 14 + 36 + 11 = 88. (8 is the next digit of the product, with 8 as a carryover to the next step).

Step 7

						A	B	C	D						
0	0	0	0	7	5	9	0	2	4	8	1		×	9	7	2	3
P	R	O	D	U	C	T					9	8	2	2	7	6	3

ABCD   Multiplier Set = 9723

5 9 0 2 ⟶ 9 7 2 3 → 5 × 3 + 9 × 2 + 0 × 7 + 2 × 9 + 8 (carryover from

Step 6) = 15 + 18 + 0 + 18 + 8 = 59. (9 is the next digit of the product, with 5 as a carryover to the next step).

Mental Mathematics

Step 8

				A	B	C	D						×	9	7	2	3
0	0	0	0	7	5	9	0	2	4	8	1		×	9	7	2	3
P	R	O	D	U	C	T				9	9	8	2	2	7	6	3

ABCD   Multiplier Set = 9723

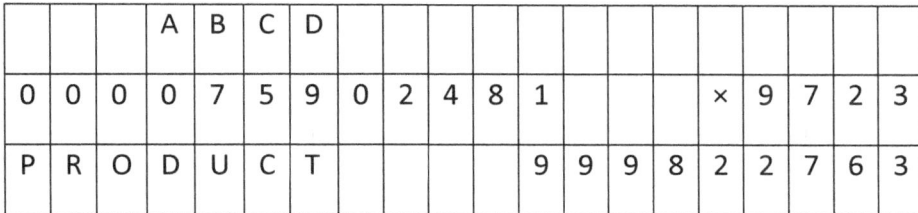

7 5 9 0 ⟶ 9723 → 7 × 3 + 5 × 2 + 9 × 7 + 0 × 9 + 5 (carryover from Step 7) = 21 + 10 + 63 + 0 + 5 = 99. (9 is the next digit of the product, with 5 as a carryover to the next step).

Step 9

				A	B	C	D							9	7	2	3
0	0	0	0	7	5	9	0	2	4	8	1		×	9	7	2	3
P	R	O	D	U	C	T			9	9	9	8	2	2	7	6	3

ABCD   Multiplier Set = 9723

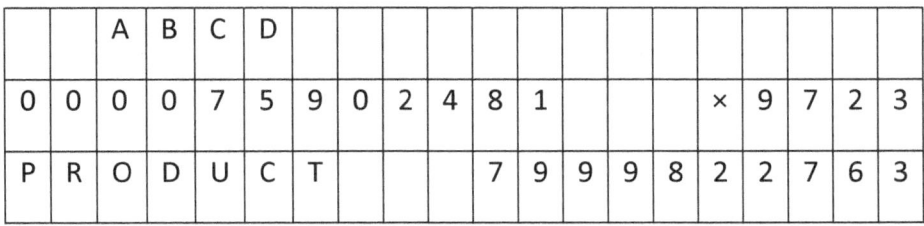

0 7 5 9 ⟶ 9723 → 0 × 3 + 7 × 2 + 5 × 7 + 9 × 9 + 9 (carryover from Step 8) = 0 + 14 + 35 + 81 + 9 = 139. (9 is the next digit of the product, with 13 as a carryover to the next step).

Step 10

				A	B	C	D							9	7	2	3
0	0	0	0	7	5	9	0	2	4	8	1		×	9	7	2	3
P	R	O	D	U	C	T			7	9	9	9	8	2	2	7	6	3

ABCD   Multiplier Set = 9723

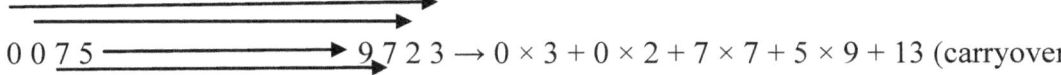

0 0 7 5 ⟶ 9723 → 0 × 3 + 0 × 2 + 7 × 7 + 5 × 9 + 13 (carryover

Mental Mathematics

from Step 9) = 0 + 0 + 49 + 45 + 13 = 107. (7 is the next digit of the product, with 10 as a carryover to the next step).

Step 11

	A	B	C	D											
0	0	0	0	7	5	9	0	2	4	8	1			×	9	7	2	3
P	R	O	D	U	C	T		3	7	9	9	9	8	2	2	7	6	3

ABCD Multiplier Set = 9723

0 0 0 7 ⟶ 9 7 2 3 →0×3 + 0×2 + 0×7 + 7×9 + 10 (carryover from Step 10) = 0 + 0 + 0 + 45 + 10 = 73. (3 is the next digit of the product, with 7 as a carryover to the next step).

Step 12

A	B	C	D												
0	0	0	0	7	5	9	0	2	4	8	1			×	9	7	2	3
P	R	O	D	U	C	T	**7**	**3**	7	9	9	9	8	2	2	7	6	3

ABCD Multiplier Set = 9723

0 0 0 0 ⟶ 9 7 2 3 ⇒ 0 × 3 + 0 × 2 + 0 × 7 + 0 × 9 + 7 (carryover from Step 11) = 0 + 0 + 0 + 0 + 7 = 7. (Digit 7 is the most significant digit of the product).

Note: The last step of our multiplication operation is, whenever the pointer ABCD is completely on the top of zeros.

Therefore, in this example step 12 is the last step, and our final product is 737999822763.

Mental Mathematics

•Multiple-Digit Numbers Multiplication- With No Modification

Now we will consider the real Trachtenberg trick for multiplication of a multi-digit number by another multi-digit number.

As we considered previously, an advantage of the Trachtenberg trick is that we can obtain a product of our multiplication directly and the same method could be used for multiplication of any two-digit number. Yet for the purpose of improvement, in many cases, it is necessary to do some modification.

Whenever we encounter multiplication with big numbers, such as 987, 769, we should add up some big numbers mentally and carry over the same big numbers to the next steps to avoid encountering such big numbers. Trachtenberg introduced the "Trick of Two Fingers" or "Units and Tens" also.

•Two Fingers Trick Procedure

1. Whenever we multiply a digit by another digit, the product is always a one-digit number or a two-digit number. $9 \times 9 = 81$ (9 is the biggest one-digit number).

2. If multiplication of two one-digit numbers produces a one digit number, then we put a zero on its left and present it as a two-digit number. $3 \times 4 = 8$, but we write it as 08, or $1 \times 9 = 9$, but we represent it as 09.

3. Every two-digit number consist of ones (units) and tens. In the number 58, digit 8 is in one's position and digit 5 is in ten's position.

⇨ $58 = 5 \times 10^1 + 8 \times 10^0$, meaning $50 + 8 = 58$.

4. In this new trick, we sometimes use ones or tens of a number. If we use ones, tens of the same number will be used in another place; therefore the result of our multiplication will not be affected.

5. As explained before, in every multiplication operation, three numbers are involved, multiplicand, multiplier, and product.

⇨ **7394 (multiplicand)** × **5842 (multiplier) = 43195748 (product).**

6. We begin our multiplication procedure by marking multiplicand digits pair by pair and multiply each pair by digits of the multiplier in a particular manner.

Mental Mathematics

7. To mark digits of the multiplicand, we use letters A and B as the pair AB, which will be placed on the top of the multiplicand and keeps moving toward left one digit at a time in each step.

8. Initially, we place B on the top of multiplication sign and assume × to be a zero, but this zero is not a part of the multiplicand and multiplier.

9. Put as many zeros on the left side of multiplicand as the number of multiplier digits.
7394 × 5842 = 43195748 will be represented as:

AB	
00007394 × 5842 =	43195748

10. Multiplication operation ends when the most significant digit of the multiplier product is zero.

Example 2 – 15:

What is the product of 8572 × 6739?

Step 1

As there are 4 digits in our multiplier, we put 4 zeros on the left hand side of the multiplicand, which is 8572 and AB is our pointer which moves one digit to the left in every step.

Pointer								A	B			
			0	0	0	0	8	5	7	2	×	6	7	3	9
Product	=											8

Multiply 9 (the least significant digit of multiplier) by digits under A and B, using two-digit formats:

9 × 2 = 18 and 9 × 0 = 00 and then we add 8 and 0 to obtain the least significant digit of the product ⇨ 8 + 0 = 8.

42

Mental Mathematics

Step 2

Now AB moves one digit to the left:

Pointer							A	B				
		0	0	0	0	8	5	7	2	×	6	7	3	9
Product	=												0	8

Multiply 9 (the least significant digit of the multiplier) by digit under A and B, using two-digit formats:

$9 \times 7 = 63$ and $9 \times 2 = 18$ and then we add 3 and 1 ⇒ 3 + 1 = 04.

1) Multiply 3 (the second digit of the multiplier from right to left) by digits under A and B, in Step 1, which are 0 and 2.
   $3 \times 2 = 06$ and $3 \times 0 = 00$ and add their last and first digits 6 + 0 = 06.
2) Add 04 and 06, to obtain the next digit of product:
   04 + 06 = 10, thus 0 is the next digit of the product, and 1 is the carryover to the next step.

Step 3

Now AB moves one digit to the left:

Pointer						A	B					
		0	0	0	0	8	5	7	2	×	6	7	3	9
Product	=											7	0	8

$9 \times 5 = 45$ and $9 \times 7 = 63$ (9 is the least significant digit of multiplier and always multiplied by digits under AB).

5 + 6 = 11.

$3 \times 7 = 21$ and $3 \times 2 = 6$.

Mental Mathematics

$1 + 0 = 1$.

$7 \times 2 = 14$ and $7 \times 0 = 00$

$4 + 0 = 4$.

$11 + 1$ the $+ 4 + 1$(carryover from Step 2) $= 17$, therefore 7 is the left adjacent digit of 0 and 1 is carryover.

Step 4

Now AB moves one digit to the left:

Pointer					A	B					
		0	0	0	8	5	7	2	×	6	7	3	9
Product	=									6	7	0	8

$9 \times 8 = 72$ and $9 \times 5 = 45$

$2 + 4 = 6$

$3 \times 5 = 15$ and $3 \times 7 = 21$

$5 + 2 = 7$.

$9 + 1 = 10$.

$7 \times 7 = 49$ and $7 \times 2 = 14$

$6 \times 2 = 12$ and $6 \times 0 = 00$

$2 + 0 = 2$.

$6 + 7 + 10 + 2 + 1 = 26$, therefore 6 is the left adjacent digit of 7, and 2 is carryover.

Step 5

Now AB moves one digit to the left:

44

Mental Mathematics

Pointer					A	B					
		0	0	0	0	8	5	7	2	×	6	7	3	9
Product	=								6	6	7	0	8

$9 \times 0 = 00$ and $9 \times 8 = 72$

$0 + 7 = 7$

$3 \times 8 = 24$ and $3 \times 5 = 15$

$4 + 1 = 5$

$7 \times 5 = 35$ and $7 \times 7 = 49$

$5 + 4 = 9.$

$6 \times 7 = 42$ and $6 \times 2 = 12$

$2 + 1 = 3.$

$7 + 5 + 9 + 3 + 2 = 26.$

Step 6

Now AB moves one digit to the left:

Pointer				A	B						
		0	0	0	0	8	5	7	2	×	6	7	3	9
Product	=							7	6	6	7	0	8

$9 \times 0 = 00$ and $9 \times 0 = 00$

$0 + 0 = 0$

$3 \times 0 = 00$ and $3 \times 8 = 24$

Mental Mathematics

$0 + 2 = 2$

$7 \times 8 = 56$ and $7 \times 5 = 35$

$6 + 3 = 9$.

$6 \times 5 = 30$ and $6 \times 7 = 42$

$0 + 4 = 4$

$0 + 2 + 9 + 4 + 2 = 17$, therefore 7 is the left adjacent digit of 6, and 1 is carryover.

Step 7

Now AB moves one digit to the left:

Pointer			A	B							
		0	0	0	0	8	5	7	2	×	6	7	3	9
Product	=						7	7	6	6	7	0	8

$9 \times 0 = 00$ and $9 \times 0 = 00$

$3 \times 0 = 00$ and $3 \times 0 = 00$

$0 + 0 = 0$.

$7 \times 0 = 00$ and $7 \times 8 = 56$

$0 + 5 = 5$.

$6 \times 8 = 48$ and $6 \times 5 = 30$

$8 + 3 = 11$.

$0 + 0 + 5 + 11 + 1 = 17$, therefore 7 is the left adjacent digit of 7, and 1 is carryover.

Step 8

Now AB moves one digit to the left:

Mental Mathematics

Pointer		A	B									
		0	0	0	0	8	5	7	2	×	6	7	3	9
Product	=					5	7	7	6	6	7	0	8

$9 \times 0 = 00$ and $9 \times 0 = 00$

$0 + 0 = 0.$

$3 \times 0 = 00$ and $3 \times 0 = 00$

$0 + 0 = 0.$

$7 \times 0 = 00$ and $7 \times 0 = 00$

$0 + 0 = 0.$

$6 \times 0 = 00$ and $6 \times 8 = 48$

$0 + 4 = 4.$

$0 + 0 + 0 + 4 + 1 = 5$, therefore 5 is the left adjacent digit of 7.

Step 9

Now AB moves one digit to the left:

Pointer		A	B									
			0	0	0	0	8	5	7	2	×	6	7	3	9
Product	=						5	7	7	6	6	7	0	8

$9 \times 0 = 00$ and $9 \times 0 = 00$

$0 + 0 = 0.$

$3 \times 0 = 00$ and $3 \times 0 = 00$

Mental Mathematics

$0 + 0 = 0$.

$7 \times 0 = 00$ and $7 \times 0 = 00$

$0 + 0 = 0$.

$6 \times 0 = 00$ and $6 \times 0 = 00$

$0 + 0 = 0$.

Therefore the final product of 8572×6739 is:

| 5 | 7 | 7 | 6 | 6 | 7 | 0 | 8 |

Mental Mathematics

Chapter 3

What is Addition?
The addition is the procedure of finding a sum or total of two or more numbers.

$5 + 17 + 128 = 150$. The numbers 5, 17, and 128 are called the addends, and 150 is the sum.

•Trachtenberg - Addition Method

In the conventional or traditional addition method, numbers are added column by column from right to left. We do the same in the Trachtenberg - Addition Method, but when the number of addends is increased, every column of our addition turns into a long column of single digit numbers and when these digits are added, they form two, three, and four-digit numbers like 10, 13, 73, and 138.

When we add these digits, the number of digits in one column starts growing bigger and bigger, causing the addition to becoming more difficult and confusing with the increase of the possibility to make a mistake, in the Trachtenberg - Addition Method, we use a certain trick to avoid increasing the number of digits in our columns, and thus, stay working with single-digit numbers only.

Example 3 – 1:

What is the sum of 487095 + 998252 + 754184 + 226190 + 59726 + 878679 + 775807 + 6439318?

1) We put all numbers in a table as shown in Table (3 – 1).
2) If the number of digits in a number is less than other numbers, we put a zero on its left-hand side, to fill the columns. (Columns N_1 to N_6 contain $digit_1$ to $digit_6$ of above numbers)
3) Columns C_1 to C_6 are counters, which will be explained later.
4) Raw Sub–Sum contains the sub-sum of columns N_1 to N_6.
5) Row Count–Sum contains the final digits for columns C_1 to C_6. Now we begin to find the sum of digits from right to left in columns N_1 to N_6. Whenever our sum is 11 or more, we increase our counter by 1. (If our counter is 0, it becomes 1, if it is 1, it becomes 2).

Mental Mathematics

6) If a sum is more than 11, we subtract 11 from the sum, and then add the remainder to the next digit in that column.
7) Once we obtain Sub–Sum and Counter–Sum, we do L addition, as shown in Example 1, to obtain the sum of our numbers.

To understand the above procedure, we are going to apply these steps in the following table:

Table 3 - 1

		N_1	C_1	N_2	C_2	N_3	C_3	N_4	C_4	N_5	C_5	N_6	C_6
	0	4		8		7		0		9		5	
	0	9	1	9	1	8	1	2		5	1	2	
	0	7		5	2	4		1		8	2	4	1
	0	2	2	2		6	2	1		9		0	
	0	0		5		9	3	7	1	2	3	6	
	0	8		7	3	8		6		7		9	2
	0	7	3	7		5	4	8	2	0		7	3
	0	6		4	4	3		9	3	1		8	
Sub - Sum	0	10		3		6		1		8		8	
Counter - Sum		3		4		4		3		3		3	
Final - Sum	4	8		2		3		8		5		1	

1) Beginning with column N_6, we have $5 + 2 + 4 = 11 \geq 11$, causing C_6 to increase by 1, thus $C_6 = 1$ (shown in front of digit 4), and then $(11 - 11) = 0$, and then we have $(0 + 6 + 9) = 15 > 11$, causing C_6 to increase by 1, thus $C_6 = 2$ (shown in front of digit 9), and then we have $(15 - 11) = 4$, and then we have $(7 + 4) = 11 \geq 11$, causing C_6 to increase by 1, thus, $C_6 = 3$, (shown in front of digit 7), and thus, 8 will be inserted in square–crossing, of sub–sum, and N_6 and 3 will be inserted in square–crossing of counter–sum and N_6.

2) Now we go to column N_5, where we have $9 + 5 = 14 > 11$, causing C_5 to increase by 1, thus $C_5 = 1$, (as shown in front of digit 5), and then $14 - 11 = 3$, which will be added to next digit in this column, thus we have $3 + 8 = 11 \geq 11$, and thus $C_5 = 2$, and then $11 - 11 = 0$, and then $9 + 2 = 11 \geq 11$, and thus $C_5 = 3$, and then $11 - 11 = 0$, and then $7 + 0 + 1 = 8$, and thus 8 will be inserted in square-crossing sub–sum and N_5, and 3 will be inserted in square counter–sum, and N_5.

3) And then by moving to N_4, we have $0 + 2 + 1 + 1 + 7 = 11 \geq 11$, causing C_4 to increase by 1, so we have $C_4 = 1$, and then $11 - 11 = 0$, and then $6 + 8 = 14 \geq 11$, so

Mental Mathematics

$C_4 = 2$, and then $14 - 11 = 3$, and then $3 + 9 = 12$, so $C_4 = 3$, and thus 1 will be inserted in square–crossing of sub–sum, and N_4 and 3 will be inserted in square–crossing of counter–sum, and N_4.

4) Now we jump into N_3 and move from top to bottom. At the top we have $(7 + 8) = 15$, which causes counter C_3 to increase by 1(shown in front of digit 8), thus $C_3 = 1$, and then $(15 - 11) = 4$, and then we have $(4 + 4 + 6) = 14 > 11$, causing C_3 to increase by 1, thus $C_3 = 2$ (shown in front of digit 6), and then $(14 - 11) = 3$, and then we have $(3 + 9) = 12 > 11$, causing C_3 to increase by 1, thus $C_3 = 3$ (shown in front of digit 9) and then we have $(12 - 11) = 1$, and then we have $(1 + 8 + 5) = 14 > 11$, causing C_3 to increase by 1, thus $C_3 = 4$, (shown in front of digit 5), $(14 - 11) = 3$, and then we have $(3 + 3) = 6$, thus 6 will be inserted in square-crossing sub–sum and N_3 and 4 will be inserted in square counter–sum and N_3.

5) Now we jump into N_2, and move from top to bottom. At the top we have $(8 + 9) = 17 > 11$, which causes counter C_2 to increase by 1(shown in front of digit 9), thus $C_2 = 1$, and then $(17 - 11) = 6$, and then we have $(6 + 5) = 11 \geq 11$, causing C_2 to increase by 1, thus $C_2 = 2$ (shown in front of digit 5), and then $(11 - 11) = 0$, and then we have $(0 + 2 + 5 + 7) = 14 > 11$, causing C_2 to increase by 1, thus $C_2 = 3$ (shown in front of digit 7), and then we have $(14 - 11) = 3$, and then we have $(3 + 7 + 4) = 14 > 11$, causing C_2 to increase by 1, thus $C_2 = 4$ (shown in front of digit 4), $(14 - 11) = 3$, and thus 3 will be inserted in square-crossing sub–sum and N_2, and 4 will be inserted in square counter–sum and N_2.

6) Now we jump into N_1, and move from top to bottom. At the top we have $(4 + 9) = 13 > 11$, which causes counter C_1 to increase by 1 (shown in front of digit 9), thus $C_1 = 1$, and then $(13 - 11) = 2$, and then we have $(2 + 7 + 2) = 11 \geq 11$, causing C_1 to increase by 1, thus $C_1 = 2$ (shown in front of digit 2), and then $(11 - 11) = 0$, and then we have $(0 + 8 + 7) = 15 > 11$, causing C_1 to increase by 1, thus $C_1 = 3$ (shown in front of digit 7), and then we have $(15 - 11) = 4$, and then we have $(4 + 6) = 10$, and thus 10 will be inserted in square-crossing sub–sum and N_1, and 3 will be inserted in square counter–sum and N_1.

7) In this step, we will determine our final–sum, digit by digit, beginning with N_6 and sub–sum square and ending with N_1 and sub–sum square, using L addition (like a knight move in chess) two squares down, and one square to the right, ignoring column C.

51

Mental Mathematics

a) Starting with **L** addition at N_6 and sub–Sum square, we have $8 + 3 + 0 = 11$, thus 1 is the least significant digit of our final–sum, and 1 will be carried over to column N_5.
b) L addition at N_5 and sub–sum square consist of $8 + 3 + 3 + 1$ (carryover) = 1**5**, thus **5** is the next digit to the left of 1, and 1of 15 is the carryover.
c) L addition at N_4 and sub–sum square consist of $1 + 3 + 3 + 1$ (carryover) = 8, thus **8** is the next digit to the left of 5.
d) L addition at N_3 and sub–sum square consist of $6 + 4 + 3 = 13$, thus **3** is the next digit to the left of 8 and 1of 13 is the carryover.
e) L addition at N_2 and sub–sum square consist of $3 + 4 + 4 + 1$ (carryover) = 12, thus digit **2** is the next digit to the left of 3 and 1 of 11 is the carryover.
f) L addition at N_1 and sub–sum square consist of $10 + 3 + 4 + 1$ (carryover) = 18, thus digit **8** is the next digit to the left of 1 and 1 of 18 is the carryover.
g) L addition at 0 column, and sub–sum square consist of $0 + 0 + 3 + 1$ (carryover) = 4, thus **4** is the next digit to the left of 8and the most significant digit of the sum.

Therefore the final–sum = 4823851.

Example 3 – 2:

What is the sum of $9815 + 2637 + 5448 + 3911 + 4726 + 3899 + 7428$ with Trachtenberg rapid arithmetic trick?

Table 3 - 2

		A	B	C	D
		9	8	1	5
		2	6	3	7
		5	4	4	8
		3	9	1	1
		4	7	2	6
		3	8	9	9
		7	4	2	8
Sub - Sum		0	2	0	0
Counter - Sum		$C_A = 3$	$C_B = 4$	$C_C = 2$	$C_D = 4$

Through Example 1 we learned how the Trachtenberg system of addition works.

Mental Mathematics

For this example, we are not going to do everything we did in Example 1 in order to try to make it shorter and shorter to prepare you to do everything mentally off the top of your head.

We begin with column D, where we have all the least significant digits (last digits), and then columns, C, B, and A (where we have the most significant digits or first digits).

D) 5, 7, 8, 1, 6, 9, 8.

1) $5 + 7 = 12 \geq 11$ thus $C_D = 1$, and then $12 - 11 = 1$ (carryover),

2) $8 + 1 + 6 + 1 = 16 \geq 11$ thus $C_D = 2$, and then $16 - 11 = 5$ (carryover),

3) $9 + 5 = 14 \geq 11$ thus $\mathbf{C_D} = 3$, and thus $14 - 11 = 3$ (carryover),

4) $8 + 3 = 11 \geq 11$ thus $\mathbf{C_D} = 4$, and thus $11 - 11 = 0$, (sub–sum) $_D = 0$.

C) 1, 3, 4, 1, 2, 9, 2

1) $1 + 3 + 4 + 1 + 2 = 11 \geq 11$ thus $C_C = 1$, and then $11 - 11 = 0$,

2) $9 + 2 + 0$ (carryover) $= 11 \geq 11$, thus $\mathbf{C_C} = 2$ and then $11 - 11 = 0$, (sub–sum) $_C = 0$.

B) 8, 6, 4, 9, 7, 8, 4.

1) $8 + 6 = 14 \geq 11$, thus $\mathbf{C_B} = 1$, and then $14 - 11 = 3$ (carryover),

2) $4 + 9 + 3 = 16 \geq 11$, thus $\mathbf{C_B} = 2$, and then $16 - 11 = 5$ (carryover),

3) $7 + 5 = 12 \geq 11$ thus $\mathbf{C_B} = 3$ and then $12 - 11 = 1$ (carryover),

4) $8 + 4 + 1 = 13 \geq 11$, thus $\mathbf{C_B} = 4$ and then $13 - 11 = 2$, (sub–sum) $_B = 2$.

A) 9, 2, 5, 3, 4, 3, and 7

1) $9 + 2 = 11 \geq 11$, thus $\mathbf{C_A} = 1$, and then $11 - 11 = 0$ (carryover),

1) $5 + 3 + 4 = 12 \geq 11$, thus $\mathbf{C_A} = 2$, and then $12 - 11 = 1$ (carryover),

1) $3 + 7 + 1$ (carryover) $= 11 \geq 11$, thus $\mathbf{C_A} = 3$, and then $11 - 11 = 0$, (sub–sum)$_A = 0$.

Mental Mathematics

Now we will do **L** addition for columns D, C, B, and A to obtain the sum.

```
 0        0        2        0        0
 4  0     2  4     4  2     3  4      0  3
```

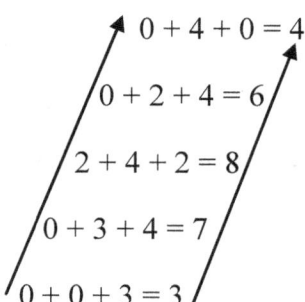

0 + 4 + 0 = 4
0 + 2 + 4 = 6
2 + 4 + 2 = 8
0 + 3 + 4 = 7
0 + 0 + 3 = 3

Sum = 37864.

•Examining Addition Correctness

To check the correctness of our sum, we will consider three factors:

1) Columns of digits
2) Rows of operation–sub–sum row and sub–counters row.
3) The result of addition or sum.

Now for each factor, we will calculate a digit with a certain rule, known as the scale of comparison, and then compare all digits obtained from each factor. If all are the same digit, we have obtained the correct sum.

At this stage, we will explain how to calculate the scale of comparison for each factor.

Step 1

To find the scale of comparison for columns of digits, we have to sum up digits of each individual column, divide it by 9 and find its remainder. If the remainder of any column is a two-digit number, we have to add these digits to obtain a single digit for each column, which is going to be the scale of comparison for that particular column.

Note: In order to make the process of division by 9 easier and faster, for every column, we will mark all 9 digits, and also the combination of digits that add up to 9,

Mental Mathematics

or a number which is divisible by 9, and then we will ignore all these numbers (assuming that they do not exist).

For example, if the digits in a column are 2, 7, 9, 1, 8, 4, 5, 7, 3, and 6, then we have, $2 + 7 = 9$, $1 + 8 = 9$, $4 + 5 = 9$, and $3 + 6 = 9$. We ignore all these digits and 9, and Therefore, we are left with 7 only, and 7 is not divisible by 9, thus 7 is the remainder, and it is the scale of comparison for this particular column.

To understand the method and procedure better, we will find the scale of comparison for Example 3 – 2, with reference to Table 3–2.

1) For column D, we have 5, 7, 8, 1, 6, 9, and 8. Here we may write $5 + 7 + 6 = 18 = 9 + 9$, and $8 + 1 = 9$, and ignoring all these digits and 9, now we are left with 8, which is not divisible by 9. Therefore, the scale of comparison for column D is 8.

2) For column C, we have 1, 3, 4, 1, 2, 9, and 2. Here we may write $1 + 3 + 4 + 1 = 9$, and ignoring all these digits and 9, we are left with $2 + 2 = 4$ which is the scale of comparison for column C.

3) For column B, we have 8, 6, 4, 9, 7, 8, and 4. Here we may write $8 + 6 + 9 = 18 = 9 + 9$, and ignoring all these digits and 9, we are left with $7 + 8 + 4 = 19$, which is $18 + 1$, and as 18 is divisible by 9; we are left with 1, which is the scale of comparison for column B.

4) For column A, we have 9, 2, 5, 3, 4, 3, and 7. Here we may write 9 and $5 + 4 = 9$, then $2 + 7 = 9$, and ignoring all these digits and 9, we are left with $3 + 3 = 6$, which is not divisible by 9, and thus 6 is the scale of comparison for column A.

Therefore the scale of comparison for columns A, B, C, and D respectively is 6, 1, 4, and 8.

⇨**The scale of comparison for columns = 6148.**

Step 2

Now we will use sub-sum and counter–sum to create another scale of comparison, and call it sub-sum and counter–sum scale of comparison.

In order to create sub-sum and counter–sum scale of comparison, we add up every digit of all columns A, B, C, and D, and whenever we encounter with a Two-Digit

Mental Mathematics

Number, we add these digits to obtain a single-digit number, as shown in Table (3 – 3), we have used counter - Sum twice.

Table 3 – 3

Sub - Sum	0	2	0	0
Counter - Sum	$C_A = 3$	$C_B = 4$	$C_C = 2$	$C_D = 4$
Repeated Counter – Sum	3	4	2	4
Column Digits - Sum	0 + 3 + 3 = 6	2 + 4 + 4 = 10	0 + 2 + 2 = 4	0 + 4 + 4 = 8
Result	6	10 ⇨ 1 + 0 = 1	4	8
Single Digit Result	6	1	4	8

Step 3

The scale of comparison for columns is 6148, and the single-digit result obtained for sub–sum and counter–sum is also 6148, and our sum is 37864.

Now we will convert each of these three numbers to a single digit:

1) 6 + 1 + 4 + 8 = 19, and then 9 + 1 = 10, and then 1 + 0 = 1
2) Our second number is the same number, and obviously its result is going to be 1.
3) But for our sum which is 37864, we have 3 + 7 + 8 + 6 + 4 = 28, and then 2 + 8 = 10, and then 1 + 0 = 1, therefore the sum we have obtained in our addition is correct.

•Add and Subtract Addition Method

Whenever you want to go to your workplace or school, you may have several options to choose a path, but all paths are not the same, and some of them are shorter than others and less time-consuming.

Most of the time we have multiple options to solve a math problem or do a mathematics operation, such as addition and multiplication, but there is one which is shorter than others that reduces the time complexity and the possibility of making mistakes.

Mental Mathematics

Here I am introducing a new rapid addition method that is not part of the Trachtenberg work, but very easy to understand.

Example 3 - 3:

Find the sum of the following numbers, with the Add and Subtract Addition Method:
7852 + 9236 + 1815 + 3957 + 8463 + 6598 + 4922.

Step 1

We will use a table to organize the procedure to make it easier to understand. The table will not be necessary as you excel at the method.

Step 2

Count the number of rows in the table containing numbers (the number of given numbers).

Step 3

Multiply the number of numbers by 5 ⇨ 7 × 5 = 35.

Step 4

Subtract 5 from each digit in columns A, B, C, and D.

Step 5

Add up the result of subtraction in columns A, B, C, and D.

Note: You may cancel symmetric digits to make the addition operation simpler.

Step 6

If the result of subtraction in a column is positive, then add it to 35; if it is negative, subtract it from 35(do Algebra addition), as shown in the Table 3 – 4.

Proceed to next page

Mental Mathematics

Table 3 – 4

NON	+4	A	+4	B	+3	C	+3	D	
1	0	7	7 – 5 = +2	8	8 – 5 = +3	5	5 – 5 = 0	2	2 – 5 = -3
2	0	9	9 – 5 = +4	2	2 – 5 = -3	3	3 – 5 = -2	6	6 – 5 = +1
3	0	1	1 – 5 = -4	8	8 – 5 = +3	1	1 – 5 = -4	5	5 – 5 = 0
4	0	3	3 – 5 = -2	9	9 – 5 = +4	5	5 – 5 = 0	7	7 – 5 = +2
5	0	8	8 – 5 = +3	4	4 – 5 = -1	6	6 – 5 = +1	3	3 – 5 = -2
6	0	6	6 – 5 = +1	5	5 – 5 = 0	9	9 – 5 = +4	8	8 – 5 = +3
7	0	4	4 – 5 = -1	9	9 – 5 = +4	2	2 – 5 = -3	2	2 – 5 = -3
	0 + 4 = 4... ...		35 + 3 + 4 = 42		35 + 10 + 3 = 48		35 + 3 + 1 -2 – 3 = 34		35 – 3 + 1 = 33
Sum = 42843	4		2		8		4		3

•Correctness check of the Basic Math Operations

In general, there are two methods to check the correctness of the basic math operations of addition, subtraction, multiplication, division, and square root.

•Digits Addition – Method

This method is also known as the Nine-Nine Method.

Step 1

Calculate the sum of the digits of given number.

Step 2

If a sum is not a single digit, repeat Step 1 until you get a single digit number.

Mental Mathematics

For example, the sum of the digits of 70513 is $7 + 0 + 5 + 1 + 3 = 16$, and then $1 + 6 = 7$.

We may do the same thing by adding $7 + 0 + 5 = 12$ and then $3 + 1 + 3 = 7$, which is easier.

Note: While adding digits of a number, we may ignore all 9 digits, and any two or three digits that add up to 9, such as $8 + 1$, $2 + 7$, or $3 + 6$, and also all zeros.

Example 3 – 4:

Find the sum of the digits in each of the following numbers:

N = 900927999936148 and M = 7459999000814322.

Answer:

If we ignore all nines and zeros in N ($2 + 7 = 9$, $3 + 5 + 1 = 9$, and $1 + 8 = 9$) we are left with 4, which is the sum of digits in number N.

In M, if we ignore all nine and zeros, ($7 + 2 = 9$, $4 + 5 = 9$, $8 + 1 = 9$, and $4 + 3 + 2 = 9$), are left with the sum of digits as zero.

Note: We may use the same rule for decimal numbers such as 39425.361 or 59.2, but in this case, we have to ignore the decimal point and then find the sum of digits.

Example 3 - 5:

Find the sum of digits in A = 39425.371.

If we ignore decimal point, and 9 ($4 + 5 = 9$, $2 + 7 = 9$), we are left with $3 + 3 + 1 = 7$, which is the sum of digits in number A.

Why is this rule correct?

Whenever we divide a number by 9, we may have a zero remainder or any digit from 1 – 8, but we may find the remainder just using the Nine-Nine Digit Addition Method.

For example, $34 \div 9 = 3$ with 7 as its remainder, but by using Nine-Nine Digit Addition Method, we have $3 + 4 = 7$, which is the same digit.

59

Mental Mathematics

Example 3 – 6:

What is the remainder of the following division operation by 9?

1) 1111100001
2) 10001111111100
3) 4990000511600
4) 9111000000230

Answer –

1×6 = 6.   2) 9×1 = 9 thus remainder is 0.   3) 1 + 1 + 6 = 8.   4) 1 + 1 + 1 + 2 + 3 = 8

• **Number Eleven–Method**
In this method, we will find the remainder of division operation of a number by eleven without doing division operation.

• **Two-Digit Numbers–Case**
In this case, we will subtract the tens digit from the units digit to find the remainder. For example, for 48, we would do 8–4 = 4.

Therefore, 11 × 4 = 44 and 44 + 4 = 48, Thus 48 ÷ 11 = 4, with 4 as its remainder

Now if we encounter with a number like 73, in which tens digit is greater than the units' digit, first we will need to add 11 to the units' digit and then we will do the subtraction operation as follows:

3 + 11 = 14, and then 14 – 7 = 7, which is the remainder of 73 ÷ 11 = 6

• **Multi-Digit Numbers–Case**
In this case, we begin from right to left and sum every other digit of a number and then subtract them from each other.

Example 3 – 7:

What is the remainder of 758920643 ÷ 11?

What is the remainder of 758920643 ÷ 11?

In order to find the sum of digits for every other digit of number 758921643, we will use a table and split the number into two different rows, as shown below in Table 3 – 5.

Mental Mathematics

Proceed to next page

Table 3 – 5

		9	8	7	6	5	4	3	2	1
The Sum Digit	M = 7 + 8 + 2 + 6 + 3 = 8	7		8		2		6		3
The Sum Digit	N = 5 + 9 + 1 + 4 = 10 = 1 + 0 = 1		5		9		1		4	
Remainder by 11	M – N = 8 – 1 = 7.									
Therefore	758921643 ÷ 11 = 68992876									
Or	758921643 = 11×68992876 + 7									
M = 7 + 8 = 15	1 + 5 = 6, 6 + 2 + 6 14, 1 + 4 + 3 = 8									
N = 5 + 9 = 14	1 + 4 = 5, 5 + 1 + 4 = 10, 1 + 0 = 1									

Note: In the above table, we have assigned a number to each digit.

M is the sum digit of digits, indicated by odd digits. N is the sum digit of digits, indicated by even numbers.

Note: To separate the digits of a number we have to move from right to left.

Chapter 4

Division

We will be using the following terminology in this section:

Dividend ÷ Divisor = Quotient

Remainder = Dividend − (Divisor × Quotient)

For example, 958741 / 39 = 24583 and 4 is the remainder of this division.

Therefore: 958741 = 24583 × 39 + 4.

In this example, 958741 is the dividend, 39 is the divisor, 24583 is quotient, and 4 is the remainder.

•The Trachtenberg Division Method

Assume 893567412 to be the dividend and 97 to be divisor, and our aim is to find the quotient and remainder.

The Trachtenberg trick avoids mistakes and is easy to perform.

Trachtenberg suggests two different methods for the division operation of numbers. The first method is not a fast method, but it is a technique which is very easy to understand, provides accuracy and avoids mistakes that may occur in traditional division method.   The second method is a faster method which requires more practice to master.

•The Trachtenberg Division Procedure

Assume M and N as the number of digits in the dividend and divisor.

Step 1

Multiply the divisor by 1-10, as shown in Table 4–1 to obtain the scales, where

Scale 0 = 0 × Divisor, Scale1 = 1 × Divisor, Scale 2 = 2 × Divisor and so on and so forth.

Mental Mathematics

Step 2

Obtain the sub1-dividend from dividend by considering the number of digits in the divisor as a base, similar to the traditional division method.

Note: Number of sub1-dividend ≥ number of divisor

Step 3

If we compare sub1–dividend with numbers in scales column, sub1–dividend falls between two numbers. Choose the number which is smaller than sub1–dividend and subtract it from sub1-dividend.

Step 4

The chosen scale, the number from Step 3, is obtained by multiplying a digit and the divisor. This digit is the first digit of our quotient (from the left-hand side).

Step 5

Put the digit from dividend which is after sub1–dividend in front of the number obtained by the subtraction operation in Step 3, obtain the sub2–dividend and the second digit of the quotient, in the same way, we did in Step 4.

Step 6

Redo the same procedure with all dividend digits to obtain the other digits of the quotient.

Once you are done with the last digit of the dividend, stop.

Example 4 – 1:

What is the quotient and remainder of $6517248749 \div 83$?

Proceed to next page Table 4 – 1.

Mental Mathematics

Table 4 - 1

Dividend	Divisor				Q	U	O	T	I	E	N	
6517248749	**83**				7	8	5	2	1	0	6	9
Sub - Dividend	Scales		Scales	(Sub – Dividend) – (Scales)	1	2	3	4	5	6	7	8
(6) 57	83×0	0	0	57 – 0 = 57						0		
(5) 88	83×1	1	83	88 – 83 = 5							1		
(4) 174	83×2	2	166	174 – 166 = 8					2			
	83×3	3	249									
	83×4	4	332									
(3) 432	83×5	5	415	432 – 415 = 17					5			
(7) 574	83×6	6	498	574 – 498 = 76							6	
(1) 651	83×7	7	581	651 – 581 = 70	7							
(2) 707	83×8	8	664	707 – 664 = 43		8						
(8) 769	83×9	9	747	769 – 747 = 22								9
	83×10	10	830									
				Remainder = 22								

Step 1

Sequence 651 is the shortest first sequence in the dividend (6517248749) divisible by the divisor (83).

Step 2

According to Scales Colum, Table 4 – 1:

664< Sequence 651 < 581 ⇨ thus we choose 581 = 83×7 ⇨Q1= 7.

Mental Mathematics

Step 3

651 − 581 = 70, now we move digit 7, the next digit of the dividend after sequence 651 to the right of 70, thus the sequence 707 will be the second sub – dividend.

Step 4

According to Scales Colum, Table 4 – 1:

664< Sequence 707 < 747⇨ thus we choose 664 = 83×8 ⇨Q_2= 8.

Step 5

707 − 664 = 43, now we move digit 2, the next digit of the dividend after digit 7 to the right of 43, thus the sequence 432 will be the third sub – dividend.

Step 6

According to Scales Colum, Table 4 – 1:

498 > Sequence 432 > 415⇨ thus we choose 415 = 83×5⇨Q_3= 5.

Step 7

432 − 415 = 17, now we move digit 4, the next digit of the dividend after digit 4 to the right of 17, thus the sequence 174 will be the fourth sub – dividend.

Step 8

According to Scales Colum, Table 4 – 1:

249 > Sequence 174 > 166⇨ thus we choose 166= 83×2⇨Q_4= 2.

Step 9

174 − 166 = 08, now we move digit 8, the next digit of the dividend after digit 4 to the right of 08, thus the sequence 088 will be the fifth sub – dividend.

Step 10

According to Scales Colum, Table 4 – 1:

166 > Sequence 88 > 83⇨ thus we choose 83= 83×1⇨Q_5= 1.

Mental Mathematics

Step 11

88 − 83 = 05, now we move digit 7, the next digit of the dividend after digit 8 to the right of 05, thus the sequence 057 will be the sixth sub – dividend.

Step 12

According to Scales Colum, Table 4 – 1:

0 > Sequence 57 > 83 ⇨ thus we choose 57 = 57×0 ⇨ Q_6 = 0.

Step 13

57 − 0 = 57, now we move digit 4, the next digit of the dividend after digit 7 to the right of 57, thus the sequence 574 will be the sixth sub – dividend.

Step 14

According to Scales Colum, Table 4 – 1:

581 > Sequence 574 > 498 ⇨ thus we choose 498 = 83×6 ⇨ Q_7 = 6.

Step 15

574 − 498 = 76, now we move digit 9, the next digit of the dividend after digit 4 to the right of 76, thus the sequence 769 will be the sixth sub – dividend.

Step 16

According to Scales Colum, Table 4 – 1:

830 > Sequence 769 > 747 ⇨ thus we choose 747 = 83×9 ⇨ Q_8 = 9.

Step 17

Remainder = 769 − 747 = 22.

•Correctness – Check

Now we will do a correctness check for this division operation with the help of The Nine–Digit Correctness Rule.

Step 1

Dividend − Remainder = 6517248749 − 22 = 6517248727.

Mental Mathematics

Step 2

Dividend − Remainder should be equal to Quotient × Divisor, therefore we need to check the correctness of the following:

6517248727 = 83 × 78521069.

Step 3

The sum digit of 6517248727 and the sum digit of 83 multiplied by 78521069 should be equal. Using the Nine-Digit Correctness Rule, we ignore all those digits that are 9 or sum up to it and zeros (5 + 4 = 9, 7 + 2 = 9, 1 + 8 = 9, and 7 + 2 = 9. We are left with 6 and 7, and the 6 + 7 = 13, and 1 + 3 = 4[6517248727 ÷ 9 = 724138747, with the remainder of 4 or 6517248727 = 724138747 × 9 = 6517248723 + 4].

Thus, the sum digit of 6517248727 is 4.

On the other side, we have 83, which is 8 + 3 = 11, and 1 + 1 = 2[(11 − 9) = 2]. We may ignore 7 + 2 = 9, 8 + 1 = 9, and 9, and we are left with 5 + 6 = 11 and 1 + 1 = 2.

Therefore, the sum digit of 78521069 × 83 is 2 × 2 = 4, which is the same as the sum digit 6517248727, and thus, our division operation is correct.

Now we will do the Eleven Numbers–Method correctness check on our division operation.

In this method, we will find the division remainders of the dividend (6517248727), the divisor (83), and the quotient by 11.
If we find the remainder of dividend to be equal to the remainder of the divisor × quotient, then we conclude that our division operation is correct.

•Rapid Division Method

In this section, we will learn how the Trachtenberg Rapid Division Method works, but before that, we need to become familiar with two new multiplication methods.

1 − U + T Multiplication Method (U stands for units digit, and T stands for tens digit).
2 − UT + T Multiplication Method (UT stands for both units and tens digit and T is just the tens digit).

Mental Mathematics

U + T Multiplication Method

Assume that we want to multiply 52 and 7 with the U & T Multiplication Method.

Table 4 – 2

	U	+	T	
52 × 7				
5×7 =	35		14	= 2×7
		5 + 1 = 6		

We use only the units digit of 35 (which is digit 5), and tens digit of 14, (which is 1) and then add 1 and 5. The result of U + T multiplication of 52 and 7 is 6.

UT + T Multiplication Method

Table 4–3

	UT	+	T	
52 × 7				
5 × 7 =	35		14	= 2×7
		35 + 1 = 36		

In this method, we use both the units digit and the tens digit of 35 which is under UT, and the tens digit of 14 which is under T.

Therefore the result of this multiplication is 35 + 1 = 36 as shown in Table 4–3.

Long Division and New Terminology

1) The length of a number is the number of its digits.

2) In any division operation, we find the shortest sequence of digits starting from the left end of the dividend, which is divisible by the divisor.

Mental Mathematics

We call this sequence, first sub – dividend, and represent it as **SD1**.

3) We represent Quotient with **Q**, and each digit of **Q**, from left to right, is represented with $Q_1, Q_2, Q_3 \ldots Q_n$, where **n** is the number of Quotient digits.

4) We represent divisor by R, and each digit of R from left to right is represented by $R_1, R_2\ldots,$ and R_m.

5) $Q_1 = SD1/R_1$

6) $SD1 = Q_1 \times R_1 + r1$, where **r1** is the remainder of $SD1/R_1$ division.

7) The pre-sub - dividend is represented by **PSD** and is a sequence moving the next digit of dividend to the right of **r**.

We will use pre-sub - dividend for calculating the second sub - dividend, the third sub - dividend...

8) If we represent dividend by **D**, and each digit of **D**, from left to right by $D_1, D_2\ldots$

Then $PSD = rD_{(N+1)}$.

We may find **SD3, SD4...** by following the same procedure.

Example 4–2:

What is the quotient and remainder of $8384 \div 32$?

Dividend	D_1	D_2	D_3	D_4			
	8	3	8	4			
Divisor	R_1	R_2					
	3	2					
Pre – Sub Dividend	PSD1	PSD2	PSD3	PSD4			
		23	08	04			
Sub – Dividend	SD1	SD2	SD3	SD4			
	8	19	6	0			
Quotient					2	6	2
Remainder					0		

Mental Mathematics

Step 1

Compare D_1 and R_1, if $D_1 \geq R_1$ then D_1 is the pre-sub - dividend and SD1 = 8.

If $R_1 > D_1$ then SD1 = D_1D_2.

$D_1 = 8$, and $R_1 = 3 \Rightarrow$ SD1= D_1= 8.

Step 2

In this step, we will obtain the first digit of our quotient (from the left side) by dividing D_1 by R_1:

$Q_1 = D_1 / R_1 = 8 / 3 = 2$, with 2 as its remainder.

Step 3

In this step, we will do UT + T multiplication of Q_1 and R_1R_2 (our divisor):

$32 \times 2 = 3 \times 2 = 06$ and $2 \times 2 = 04 \Rightarrow$ (UT + T) $_{Q1}$ = 06 + 0 = 6

Step 4

In this step, we will subtract UT + T) $_{Q1}$ from D_1.

D_1 - (UT + T) $_{Q1}$ = 8 − 6 = 2.

Step 5

In this step, we will find the pre–sub dividend, which is D_1 - (UT + T) $_{Q1}$ D_2 = 23.

$\Rightarrow$ PSD2 = 23.

Step 6

In this step, we will do U multiplication of Q_1, and R_1R_2:

$32 \times 2 = 2 \times 2 = 04$.

Step 7

In this step, we will obtain sub dividend (SD2), by subtracting U multiplication of Q_1 and R_1R_2 from PSD2, $\Rightarrow$ SD2 = 23 − 4 = 19.

Mental Mathematics

Step 8

In this step, we will obtain Q_2, the second digit of our quotient:

$Q_2 = SD2 / R_1 = 19 / 3 = 6$, with 1 as its remainder.

Step 9

In this step, we will do UT + T multiplication of Q_2 and R_1R_2:

$32 \times 6 = 3 \times 6 = 18$ and $2 \times 6 = 12$, and then $18 + 1 = 19$, and now we will subtract this result from SD2 ⇨ $19 - 19 = 0$.

Step 10

In this step, we will obtain the PSD3, using the result of Step 9, and D_3

⇨ $PSD3 = SD2D_3 = 08$.

Step 11

In this step, we will perform U multiplication of Q_2 and R_1R_2:

$32 \times 6 = 2 \times 6 = 12 =$ ⇨ U multiplication of Q_2 and R_1R_2 is 2.

Subtract this result from PSD3 to obtain SD3 ⇨ $SD3 = 8 - 2 = 6$.

Step 12

In this step, we will obtain Q_3:

$Q_3 = SD3 / R_1 = 6 / 3 = 2$.

Step 13

In this step, we will do UT + T multiplication of Q_3 and R_1R_2:

$32 \times 2 = 3 \times 2 = 06$ and $2 \times 2 = 04$, and then $06 + 0 = 6$

Subtract this result from SD3 ⇨ $6 - 6 = 0$.

Step 14

In this step, we will obtain PSD4, using the result of Step 13 and D_4

Mental Mathematics

⇨ PSD4 = SD3D$_4$ = 04.

Step 15

In this step, we will perform U multiplication of Q$_3$ and R$_1$R$_2$:

32 × 2 = 2 × 2 = 4, and subtract this result from PSD4 to obtain SD4.

⇨ SD4 = 4 − 4 = 0.

The result of zero here indicates that there is no remainder for this division operation. Therefore, at this point, we are done with our division operation.

This procedure becomes much faster once you master the technic and with some practice, you may do it even in less than a minute.

Example 4 – 3:

What is the quotient and remainder of 1615 ÷ 31?

Step 1

R$_1$ > D$_1$ thus SD1 = D$_1$D$_2$ = 16, and thus Q$_1$ = SD1 / R$_1$ = 16 / 3 = 5.

Dividend	D$_1$	D$_2$	D$_3$	D$_4$	
	1	6	1	5	
Divisor	R$_1$	R$_2$			
	3	1			
Pre – Sub Dividend	PSD1	PSD2	PSD3	PSD4	
		11			
Sub Dividend	SD1	SD2	SD3	SD4	
	16	8			
Quotient	Q$_1$	Q$_2$	Q$_3$	Q$_4$	
	5	2			
Remainder	3				

Step 2

Perform UT + T multiplication of Q$_1$ on R$_1$R$_2$ ⇨ 31 × 5 = 3 × 5 = 15 and 1 × 5 = 05

⇨ 15 + 0 = 15

Mental Mathematics

PSD2 = [SD1 − (UT + T)] D_3 = [16 − 15]1 = 11.

U multiplication of R_1R_2 on Q_1 ⇨ 31 × 5 = 1 × 5= 5

SD2 = PSD2 − U multiplication on Q_1 = 11 − 5 = 6 ⇨ Q_2 = SD2 / R_1 = 6 / 3 = 2.

Step 3

UT + T multiplication on Q_2 = 31 × 2 = 3 × 2=06 and 1 × 2= 02 ⇨ 06 + 0= 6.

PSD3 = [SD2 − (UT + T)] D_4 = [6 − 6]5 = 05.

U multiplication on Q_2 = 31 × 2 = 2.

SD3 = PSD3 − U multiplication on Q_2= 5 − 2 = 3.

As there is no more digit in the dividend, this is the end of our division operation, and the SD3 value is the remainder of the division operation, ⇨ Remainder = SD3 = 3.

Example 4 − 4:

What is the quotient and remainder of 2294 ÷ 62?

Step 1

R_1 > D_1 thus SD1 = D_1D_2 = 22 ⇨ Q_1 = SD1 / R_1 = 22 / 6 = 3.

Dividend	D_1	D_2	D_3	D_4
	2	2	9	4
Divisor	R1	R2		
	6	2		
Pre − Sub Dividend	PSD1	PSD2	PSD3	PSD4
		49	04	
Sub Dividend	SD1	SD2	SD3	SD4
	22	48	0	
Quotient	Q_1	Q_2	Q_3	Q_4
	3	7		
Remainder	0			

Step 2

UT + T multiplication on Q_1 = 62 × 3 ⇨ 6 × 3= 18 and 2 × 3 =06 ⇨ 18 + 0 = 18.

Mental Mathematics

PSD2 = [SD1 − (UT + T)] D$_3$ = [22 − 18]9 = 49.

U multiplication on Q$_1$ = 62 × 3 ⇨ 2 × 3 = 6.

SD2 = PSD2 − U multiplication on Q$_1$ = 49 − 6 = 43, ⇨ Q$_2$ = SD2 / R1 = 43 / 6 = 7.

Step 3

UT + T multiplication on Q$_2$ = 62 × 7 = 6 × 7 = 42 and 2 × 7 = 14 ⇨ 42 + 1 = 43.

PSD3 = [SD2 − (UT + T)] D$_4$ = [43 − 43]4 = 04.

U multiplication on Q$_2$ = 62 × 7 = 14 ⇨ U multiplication on Q$_2$ = 4

SD3 = PSD3 − U multiplication on Q$_2$ = 4 − 4 = 0.

D$_4$ = 4 is the last digit of dividend we used and SD3 = 0 too, therefore we are done with our division operation, and the remainder of this division is also zero.

Note: Whenever the second digit of the divisor is 8 or 9, we do division operation, after adding a 1 to the first digit of the divisor.

Example 4 − 5:

Perform the following division operation, 2028 ÷ 39.

Step 1

R$_2$ = 9, thus we add 1 to R$_1$, and do division on (R$_1$ + 1) = 3 + 1 = 4.

(R$_1$ + 1) = 4 > D$_1$ (2), thus SD1 = D$_1$D$_2$ = 20 ⇨ Q$_1$ = SD1 / (R$_1$ + 1) = 20 / 4 = 5.

Step 2

UT + T multiplication on Q$_1$ = 39 × 5 = 3 × 5 = 15 and 9 × 5 = 45) ⇨ 15 + 4 = 19.

PSD2 = [DS1 − (UT + T)] D$_3$ = [20 − 19]2 = 12.

U multiplication on Q$_1$ = 39 × 5 = 45 ⇨ U multiplication on Q$_1$ = 5.

SD2 = PSD2 - U multiplication on Q$_1$ = 12 − 5 = 7 ⇨ Q$_2$ = SD2 / R1 = 7 / 3 = 2.

Mental Mathematics

Step 3

UT + T multiplication on $Q_2 = 39 \times 2 = 3 \times 2 = 6$ and $9 \times 2 = 18 \Rightarrow 6 + 1 = 7$.

PSD3 = [SD2 − (UT + T)] D_4 = [7 − 7]8 = 08.

U multiplication on $Q_2 = 39 \times 2 = 9 \times 2 = 18$) ⇒ U multiplication on $Q_2 = 8$.

SD3 = PSD3 − U multiplication on $Q_2 = 8 − 8 = 0$, therefore the remainder of our division is zero.

Dividend	D_1	D_2	D_3	D_4	
	2	0	2	8	
Divisor	R_1	R_2			
	3	9			
PSD	PSD1	PSD2	PSD3	PSD4	
		12	08		
SD	SD1	SD2	SD3	SD4	
	20	7	0		
UT + T multiplication on Q	U - Q_1	U − Q_2			
	19	8			
U multiplication on Q					
Remainder					0
Quotient	Q_1	Q_2			Q
	5	2			52

•Division Operation with Three-Digit Divisor

This is very similar to the division of two-digit divisors we already learned, with the difference being a third digit in the divisor.

Example 4 − 6:

What is the quotient and remainder of the following division operation: 236831 ÷ 674?

Step 1

$R_1 > D_1$ thus SD1 = D_1D_2 = 23 and therefore Q_1 = SD1 / R_1 = 23 / 6 = 3.

Mental Mathematics

Step 2

Perform UT + T multiplication of Q_1 on R_1R_2.

⇨ $67 \times 3 = 6 \times 3 = 18$ and $7 \times 3 = 21$ ⇨ $18 + 2 = 20$.

PSD2 = [SD1 − (UT + T)] D_3 = [23 − 20]6 = 36.

Perform U + T multiplication of Q_1 on R_2R_3.

$74 \times 3 = 7 \times 3 = 21$ and $4 \times 3 = 12$ ⇨ $1 + 1 = 2$

SD2 = [(PSD2 − (U + T)] = 36 − 2 = 34 ⇨ Q2 = SD2 / R_1 = 34 / 6 = 5.

Step 3

Perform UT + T multiplication of Q_2 on R_1R_2.

$67 \times 5 = 6 \times 5 = 30$ and $7 \times 5 = 35$ ⇨ $30 + 3 = 33$

PSD3 = [SD2 − (UT + T)] D_4 = (34 − 33)8 = 18.

Perform U + T multiplication of Q_2 on R_2R_3 = $74 \times 5 = 7 \times 5 = 35$ and $4 \times 5 = 20$

⇨Perform U + T multiplication of Q_2 on R_2R_3 = $5 + 2 = 7$.

Perform U multiplication of Q_1 on R_3 = $4 \times 3 = 12$ ⇨ U multiplication of Q_1 on R_3 = 2.

SD3 = [PSD3 − (U + T multiplication of Q_2 on R_2R_3) − (U multiplication of Q_1 on R_3)]

⇨SD3 = PSD3 − (7 + 2) = 18 − 9 = 9 ⇨ Q_3 = 9 / 6 = 1.

Dividend	D_1	D_2	D_3	D_4	D_5	D_6
	2	3	6	8	3	1
Divisor	R_1	R_2	R_3			
	6	7	4			
Quotient	Q_1	Q_2	Q_3			
	3	5	1			
Remainder	2	5	7			

Mental Mathematics

Example 4 – 6:

What is the quotient and remainder of 63123257 ÷ 983?

For clarity, each step is blocked separately, and you may use the same procedure for other problems in this chapter as well.

Dividend		D_1	D_2	D_3	D_4	D_5	D_6	D_7	D_8
		6	3	1	2	3	2	5	7
Divisor		R_1	R_2	R_3					
		9	8	3					
Pre – Sub Dividend		PSD_1	PSD_2	PSD_3	PSD_4	PSD_5	PSD_6		
Sub Dividend		SD_1	SD_2	SD_3	SD_4	SD_5	SD_6		
Quotient		Q_1	Q_2	Q_3	Q_4	Q_5			
Quotient		6	4	2	1	4			
Remainder = 895									

Each step of Example 4-6 is a block which shows the sub-steps to follow in order to find each of the quotient digits and also the remainder of the division operation.

Step 1

$R_1 > D_1 \Rightarrow SD1 = 63$ and $R_2 = 8 \Rightarrow Q_1 = (SD1) / (R_1 + 1) = 63/ (9 + 1) = 63/10 = 6$.

Step 2

$(UT + T) = R_1 R_2 \times Q_1 = 98 \times 6 \Rightarrow 9 \times 6 = 54$ and

$8 \times 6 = 48 = 54 + 4 = 58$.

$PSD2 = [SD1 - (UT + T)] D_3 = [63 - 58]1 = 51$.

$(U + T) = R_2 R_3 \times Q_1 = 83 \times 6 \Rightarrow 8 \times 6 = 48$ and

$3 \times 6 = 18 \Rightarrow (U + T) = 8 + 1 = 9$.

$SD2 = [PSD2 - (U + T)] = 51 - 9 = 42$.

$\Rightarrow Q_2 = SD2 / (R_1 + 1) = 42/ (9 + 1) = 4$.

Mental Mathematics

Step 3

$(UT + T) = R_1R_2 \times Q_2 = 98 \times 4 \Rightarrow 9\times4 = 36$ and

$8\times4 = 32 \Rightarrow 36 + 3 = 39$

$PSD3 = [SD2 - UT + T] D_4 = [42 - 39]2 = 32.$

$(U + T) = R_2R_3 \times Q_2 = 83 \times 4 \Rightarrow 8\times4 = 32$ and

$U = 3\times4 = 12 \Rightarrow (U + T) = 1 + 2 = 3.$

$SD3 = [PSD3 - [(U = T) + U] = [32 - (3 + 8)] = 21.$

$\Rightarrow Q_3 = SD3/ (R_1 + 1) = 21/10 = 2.$

Proceed to next page.

Step 4

$(UT + T) = R_1R_2 \times Q_3 = 98 \times 2 \Rightarrow 9\times2 = 18$ and

$8\times2 = 16 \Rightarrow (UT + T) = 18 + 1 = 19.$

$PSD4 = [SD3 - UT + T] D_5 = [21 - 19]3 = 23.$

$(U + T) = R_2R_3 \times Q_3 = 83 \times 2 \Rightarrow 8\times2 = 16$ and

$3\times2 = 6 \Rightarrow (U + T) 6 + 0 = 6$

$U = R_3 \times Q_2 = 3 \times 4 = 12 \Rightarrow U = 2.$

$SD4 = [PSD4 - ((U + T) + U] = 23 - (6 + 2) = 15.$

$\Rightarrow Q_4 = SD4/ (R_1 + 1) = 15/10 = 1.$

Mental Mathematics

Step 5

> $(UT + T) = R_1R_2 \times Q_4 = 98 \times 1 = 09$ and $08 = 09 + 0 = 9$.
>
> $PSD5 = [SD4 - (UT + T)] D_6 = [15 - 9]2 = 62$.
>
> $(U + T) = R_2R_3 \times Q_4 = 83 \times 1 = 08$ and $01 = 8 + 0 = 8$.
>
> $U = R_3 \times Q_3 = 3 \times 2 = 06 = 6$.
>
> $SD5 = [PSD5 - ((U + T) + U)] = [62 - ((8) + 6)] = 48$.
>
> $Q_5 = SD5/ (R_1 + 1) = 48/10 = 4$.

Quotient = $Q_1Q_2Q_3Q_4Q_5$ = 64214.

Step – 6

Remainder Determination

To determine the remainder of the division operation, we will continue the same way we determined subquotient, with some modification, as follows:

> $(UT + U) = R_1R_2 \times Q_5 = 98 \times 4 = 36$ and $32 = 36 + 3 = 39$.
>
> $PSD6 = [SD5 - (UT + T)] D_7 = [48 - 39]5 = 95$.
>
> $(U + T) = R_2R_3 \times Q_5 = 83 \times 4 = 32$ and $12 = 2 + 1 = 3$.
>
> $U = R_3 \times Q_4 = 3 \times 1 = 03 = 3$.
>
> $SD6 = PSD6 - [(U + T) + U] = 95 - [3 + 3] = 89$.
>
> Now we move D_8 (7) to the right of 89, to obtain 897, and then determine $R_3 \times Q_5$ result with U multiplication rule as follows:
>
> $R_3 \times Q_5 = 3 \times 4 = 12$ and final result is just 2.
>
> Now we subtract $R_3 \times Q_5$ (2) from $SD6D_8$ (897) to determine the remainder.

Mental Mathematics

Remainder = 897 − 2 = 895.

Example 4 – 7:

What is the quotient and remainder of 39863907 ÷ 729?

Dividend	D_1	D_2	D_3	D_4	D_5	D_6	D_7	D_8
	3	9	8	6	3	9	2	8
Divisor	R_1	R_2	R_3					
	7	2	9					
Pre-Sub-Dividend	PSD1	PSD2	PSD3	PSD4	PSD5	PSD6	PSD7	PSD8
Sub-Dividend	SD1	SD2	SD3	SD4	SD5	SD6	SD7	SD8
Quotient	Q_1	Q_2	Q_3	Q_4	Q_5	Q_6	Q_7	Q_8
	5	4	6	8	3			

Step 1

$D_1 < R_1 \Rightarrow SD1 = D_1D_2 = 39 \Rightarrow$

$Q1 = SD1 / R_1 = 39 / 7 = 5 \Rightarrow Q1 = 5$

Step 2

1	$(UT + T) = R_1R_2 \times Q_1 = 72 \times 5 \Rightarrow 7 \times 5 = 35$ and $2 \times 5 = 10 \Rightarrow (UT + T) = 35 + 1 = 36$
2	$PSD2 = [SD1 - (UT + T)]D_3 = [39 - 36]8 = 38$
3	$(U + T) = R_2R_3 \times Q_1 = 29 \times 5 \Rightarrow 2 \times 5 = 10$ and $9 \times 5 = 45 \Rightarrow (U + T) = 0 + 4 = 4$
4	$SD2 = PSD2 - (U + T) = 38 - 4 = 34$
5	$Q_2 = SD2 / R_1 = 34 / 7 = 4$

Proceed to next page.

Mental Mathematics

Step 3

1	$(UT + T) = R_1R_2 \times Q_2 = 72 \times 4 \Rightarrow 7 \times 4 = 28$ and $2 \times 4 = 08 \Rightarrow (UT + T) = 28 + 0 = 28$
2	$PSD3 = [SD2 - (UT + T)]D_4 = [34 - 28]6 = 66$
3	$(U + T) = R_2R_3 \times Q_2 = 29 \times 4 \Rightarrow 2 \times 4 = 08$ and $9 \times 4 = 36 \Rightarrow (U + T) = 8 + 3 = 11$
4	$U = R_3 \times Q_1 = 9 \times 5 = 45 \Rightarrow U = 5$
5	$SD3 = [PSD3 - (U + T) - U] = 66 - 11 - 5 = 50$
6	$Q3 = SD3 / R_1 = 50 / 7 = 7$. If we go to Step 4, we realize that $Q3 = 7$, $(UT + T) = 72*7 = 49$ and 14 and thus $UT + T = 49 + 1 = 50$, but $UT + T$ should always be less than SD. Therefore we do an adjustment here, and decrease Q_3 by 1, and $\Rightarrow Q_3 = 7 - 1 = 6$.

Step 4

1	$(UT + T) = R_1R_2 \times Q_3 = 72 \times 6 \Rightarrow 7 \times 6 = 42$ and $2 \times 6 = 12 \Rightarrow (UT + T) = 42 + 1 = 43$
2	$PSD4 = [SD3 - (UT + T)] D_5 = [50 - 43]3 = 73$
3	$(U + T) = R_2R_3 \times Q_3 = 29 \times 6 \Rightarrow 2 \times 6 = 12$ and $9 \times 6 = 54 \Rightarrow (U + T) = 2 + 5 = 7$
4	$U = R_3 \times Q_2 = 9 \times 4 = 36 \Rightarrow U = 6$
5	$SD4 = [PSD4 - (U + T) - U] = 73 - 7 - 6 = 60$
6	$Q_4 = SD4 / R_1 = 60 / 7 = 8$

Step 5

1	$UT + T = R_1R_2 \times Q_4 = 72 \times 8 \Rightarrow 7 \times 8 = 56$ and $2 \times 8 = 16 \Rightarrow (UT + T) = 56 + 1 = 57$
2	$PSD5 = [SD4 - (UT + T)] D_6 = [60 - 57]9 = 39$
3	$U + T = R_2R_3 \times Q_4 = 29 \times 8 \Rightarrow 2 \times 8 = 16$ and $9 \times 8 = 72 \Rightarrow (U + T) = 6 + 7 = 13$
4	$U = R_3 \times Q_3 = 9 \times 6 = 54 \Rightarrow U = 4$
5	$SD5 = [PSD5 - (U + T) - U] = 39 - 13 - 4 = 22$
6	$Q_5 = SD5 / R_1 = 22 / 7 = 3$

As already explained, if the number of digits in the dividend is N and the number of digits in the divisor is M, to obtain the quotient, we use only [N - (M - 1)] digits of the dividend. In this example, we have 8 digits in the dividend, and 3 digits in the divisor, therefore N = 8, and M = 3, thus [N - (M - 1)] = 8 - (3 - 1)] = 6.

Thus, D_6 is the last digit of the dividend to be used for obtaining the quotient, and D_7 and D_8 are used to calculate the remainder.

Mental Mathematics

Step 6

Remainder Determination

To calculate the remainder of the division operation, we will continue the same method we used in previous steps, with minor changes.

1	$UT + T = R_1R_2 \times Q_5 = 72 \times 3 \Rightarrow 7\times3 = 21$ and $2\times3 = 06 \Rightarrow 21 + 0 = 21$
2	$PSD6 = [SD5 - (UT + T)] D_7 = [22 - 21]2 = 12$
3	$U + T = R_2R_3 \times Q_5 = 29 \times 3 \Rightarrow 2\times3 = 6$ and $9\times3 = 27 \Rightarrow 6 + 2 = 8$
4	$U = R_3 \times Q_4 = 9 \times 8 = 72 \Rightarrow U = 2$
5	$SD6 = [PSD6 - (U + T) - U] = 12 - 8 - 2 = 2$
6	At this point, we do a minor modification, and in order to obtain the PSD7, we insert D_8 to the right of SD6, and then to obtain the remainder, we subtract the U multiplication of R_3 and Q_5 from PSD7, as follows: $PSD7 = SD6D_8 = 28$ and $R_3 \times Q_5 = 9 \times 3 = 27 \Rightarrow R3 \times Q5 = 7$.
7	$\Rightarrow$ the remainder $= [SD6D_8 - R_3 \times Q_5] = 28 - 7 = 21$

Example 4 - 8:

What is the quotient and remainder of $13671514 \div 4217$?

Dividend		D_1	D_2	D_3	D_4	D_5	D_6	D_7	D_8
		1	3	6	7	1	5	1	4
Divisor		R_1	R_2	R_3	R_4				
		4	2	1	7				
Pre-Sub-Dividend		PSD1	PSD2	PSD3	PSD4	PSD5	PSD6	PSD7	PSD8
			16	27	21				
Sub-Dividend		SD1	SD2	SD3	SD4	SD5	SD6	SD7	SD8
		13	10	18	9				
Quotient		Q_1	Q_2	Q_3	Q_4	Q_5	Q_6	Q_7	Q_8
		3	2	4	2				
Remainder	0								

Step 1

$D_1 < R_1 \Rightarrow SD1 = 13$ and $Q_1 = SD1 / R_1 = 13 / 4 = 3 \Rightarrow Q_1 = 3$

Proceed to next page.

Step 2

1	$(UT + T)_{Q1} = R_1R_2 \times Q_1 = 42 \times 3 \Rightarrow 4\times3 = 12$ and $2\times3 = 6 \Rightarrow (UT + T)_{Q1} = 12 + 0 = 12$
2	$PSD2 = [SD1 - (UT + T)]D_3 = [13 - 12]6 = 16$
3	$(U + T)_{Q1} = R_2R_3 \times Q_1 = 21 \times 3 \Rightarrow 2\times3 = 6$ and $1\times3 = 03 \Rightarrow (U + T)_{Q1} = 6 + 0 = 6$
4	$SD2 = PSD2 - (U + T)_{Q1} = 16 - 6 = 10$
5	$Q_2 = SD2 / R_1 = 10 / 4 = 2$

Step 3

1	$(UT + T)_{Q2} = R_1R_2 \times Q_2 = 42 \times 2 \Rightarrow 4\times2 = 8$ and $2\times2 = 4 \Rightarrow (UT + T)_{Q2} = 8 + 0 = 8$
2	$PSD3 = [SD2 - (UT + T)_{Q2}]D_4 \Rightarrow PSD3 = [10 - 8]7 = 27$
3	$(U + T)_{Q2} = R_2R_3 \times Q_2 = 21 \times 2 \Rightarrow 2\times2 = 04$ and $1\times2 = 02 \Rightarrow (U + T)_{Q2} = 4 + 0 = 4$
4	$(U + T)_{Q134} = R_3R_4 \times Q_1 = 17 \times 3 \Rightarrow 1\times3 = 3$ and $7\times3 = 21 \Rightarrow (U + T)_{Q134} = 3 + 2 = 5$
5	$SD3 = [PSD3 - (U + T)_{Q2} - (U + T)_{Q134}] = 27 - 4 - 5 = 18$
6	$Q_3 = SD3 / R_1 = 18 / 4 = 4$

Proceed to next page.

Step 4

1	$(UT + T)_{Q3} = R_1R_2 \times Q_3 = 42 \times 4 \Rightarrow 4\times4 = 16$ and $2\times4 = 08 \Rightarrow (UT + T)_{Q3} = 16 + 0 = 16$
2	$PSD4 = [SD3 - (UT + T)_{Q3}]D_5 = [18 - 16]1 = 21$
3	$(U + T)_{Q3} = R_2R_3 \times Q_3 = 21 \times 4 = 08$ and $04 = 08 + 0 = 8$
4	$(U + T)_{Q234} = R_3R_4 \times Q_2 = 17 \times 2 \Rightarrow 1\times2 = 02$ and $7\times2 = 14 \Rightarrow (U + T)_{Q2} = 2 + 1 = 3$
5	$U_{Q1} = R_4 \times Q_1 = 7 \times 3 = 21$ or 1
6	$SD4 = [PSD4 - (U + T)_{Q3} - (U + T)_{Q234} - U_{Q1}] = 21 - 8 - 3 - 1 \Rightarrow SD4 = 9$
7	$Q_4 = SD4 / R_1 = 9 / 4 = 2$

Step 5

Remainder Determination

To determine the remainder of the division operation, we will continue the same way as in previous steps, with some minor changes.

Proceed to next page.

Mental Mathematics

#	
1	$(UT + T)_{Q4} = R_1R_2 \times Q_4 = 42 \times 2 \Rightarrow 4\times 2 = 08$ and $2\times 2 = 04 = 8 + 0 = 8$
2	$PSD5 = [SD4 - (UT + T)_{Q4}] D_6 = [9 - 8]5 = 15$
3	$(U + T)_{Q4} = R_2R_3 \times Q_4 = 21 \times 2 \Rightarrow 2\times 2 = 04$ and $1\times 2 = 02 \Rightarrow (U + T)_{Q4} = 4 + 0 = 4$
4	$(U + T)_{Q334} = R_3R_4 \times Q_3 = 17 \times 4 \Rightarrow 1\times 4 = 04$ and $7\times 4 = 28 \Rightarrow (U + T)_{Q334} = 4 + 2 = 6$
5	$U_{Q2} = R_4 \times Q_2 = 7 \times 2 = 14$ thus $U_{Q2} = 4$
6	$SD5 = [PSD5 - (U + T)_{Q4} - (U + T)_{Q3} - U_{Q2}] = 15 - 4 - 6 - 4 = 1$
7	At this point we do a minor modification. In order to obtain the PSD6, we will insert D_7 to the right of SD5, thus $PSD6 = SD5D_7 = 11$. To obtain the SD6, first we find $U + T$ for $R_3R_4 \times Q_4 = 17 \times 2 = 02$ and $14 = 2 + 1 = 3$. Then we find U for $R_4 \times Q_3 = 7 \times 4 = 28 \Rightarrow U_{R4Q3} = 8$. $SD6 = PSD6 - (U + T)_{R3R4\,Q4} - U_{R4Q3} = 11 - 3 - 8 = 0$.
8	$PSD7 = SD6D_8 = 04$, but $D_8 = 4$ is the last digit of the dividend, thus we have to obtain the U for $Q_4 \times R_4 \Rightarrow U_{Q4R3} = 7 \times 2 = 14 \Rightarrow U_{Q4R3} = 4$, and now we will subtract it from PSD7 to obtain the remainder: Remainder $= PSD7 - U_{Q4R3} == 4 - 4 = 0$.

Long Division

For every long division operation such as Example 4–8, we have to follow a number of steps, and each step includes a certain number of instructions used repeatedly in a consecutive order.

Step 1

In this step, we will compare D_1 and R_1 (the first digit of the dividend with the first digit of the divisor). If R_1 is lesser or equal to D_1, then we take D_1 as the first sub–dividend. Otherwise, D_1D_2 is the first sub–dividend (SD1).

Now we will obtain Q_1 (the first digit of the quotient) by dividing D_1 or D_1D_2 by R_1 (SD1 = D_1 or SD1 = D_1D_2) $\Rightarrow Q_1 = SD1 / R_1$.

Step 2

Instruction1
In every step, we do $(UT + T)$ multiplication of the first two digits of the divisor by the newest digit of the quotient.

Obtain the pre-sub-dividend (PSD) using the following formula:

$$PSD_N = [SD_{(N-1)} - (UT + T)] D_N$$

Mental Mathematics

Instruction 2

Do (U + T) multiplication of the newest digit of the quotient by the second and third digits of the divisor

Instruction 3

Then we move onward and do (U + T) multiplication of previous quotient digits by other digits of the divisor pair by pair, and then do U multiplication for the last digit of the divisor.

Trachtenberg called this the incomplete (U + T) multiplication.

Example 4–9:

What is the quotient and remainder of 13671514 ÷ 4217?

D_1	D_2	D_3	D_4	D_5	D_6	D_7	D_8	R_1	R_2	R_3	R_4	Q_1	Q_2	Q_3	Q_4
1	3	6	7	1	5	1	4	4	2	1	7	3	2	4	2

Step 1

$R_1 > D_1 \Rightarrow SD1 = D_1D_2 = 13 \Rightarrow Q_1 = SD1 / R_1 = 13 / 4 = 3$.

Step 2

Instruction 1

$(UT + T)_{Q1} = R_1R_2 \times Q_1 = 42 \times 3 \Rightarrow 4 \times 3 = 12$ and $2 \times 3 = 06$

$\Rightarrow (UT + T)_{Q1} = 12 + 0 = 12$.

"We do a (UT + T) multiplication on R_1R_2, for every new Q we obtain".

$PSD2 = [SD1 - (UT + T)_{Q1}] D_3 = [13 - 12]6 = 16$.

Instruction 2

$(U + T)_{Q123} = R_2R_3 \times Q_1 = 21 \times 3 \Rightarrow 2 \times 3 = 06$ and $1 \times 3 = 03 = 06 + 0 = 6$.

$SD2 = [PSD2 - (U + T)_{Q123}] = 16 - 6 = 10 \Rightarrow Q_2 = SD2 / R_1 = 10 / 4 = 2$.

In this, example there is no digit in the quotient before Q_1, thus instruction 3, is not applicable here.

Mental Mathematics

Step 3

Instruction 1

$(UT + T)_{Q2} = R_1R_2 \times Q_2 = 42 \times 2 \Rightarrow 4\times2 = 08$ and $2\times2 = 04$

$\Rightarrow (UT + T)_{Q2} = 08 + 0 = 8.$

$PSD3 = [SD2 - (UT + T)_{Q2}] D_4 = [10 - 8]7 = 27.$

Instruction 2 and Instruction 3

$(U + T)_{Q223} = R_2R_3 \times Q_2 = 21 \times 2 \Rightarrow 2\times2 = 04$ and $1\times2 = 02$

$\Rightarrow (U + T)_{Q223} = 04 + 0 = 4.$

$(U + T)_{Q134} = R_3R_4 \times Q_1 = 17 \times 3 \Rightarrow 1\times3 = 03$ and $7\times3 = 21$

$\Rightarrow (U + T)_{Q134} = 03 + 2 = 5.$

$SD3 = [PSD3 - (U + T)_{Q223} - (U + T)_{Q134}] = 27 - 4 - 5$

$\Rightarrow Q_3 = SD3 / R_1 = 18 / 4 = 4.$

Step 4

In this step, we will calculate the Q_4 last digit of the quotient, and three other digits of the dividend are used to find the remainder of the division operation.

Instruction 1

$(UT + T)_{Q3} = R_1R_2 \times Q_3 = 42 \times 4 \Rightarrow 4\times4 = 16$ and $2\times4 = 08$

$\Rightarrow (UT + T)_{Q3} = 16 + 0 = 16.$

$PSD4 = [SD3 - (UT + T)_{Q3}] D_5 = [18 - 16]1 = 21.$

Instruction 2 and Instruction 3

$(U + T)_{Q323} = R_2R_3 \times Q_3 = 21 \times 4 \Rightarrow 2\times4 = 08$ and $1\times4 = 04 \Rightarrow (U + T)_{Q323} = 8 + 0 = 8.$

$(U + T)_{Q234} = R_3R_4 \times Q_2 = 17 \times 2 \Rightarrow 1\times2 = 02$ and $7\times2 = 14$

$\Rightarrow (U + T)_{Q234} = 02 + 1 = 3.$

Mental Mathematics

$U_{Q1} = R_4 \times Q_1 = 7 \times 3 = 21 \Rightarrow U_{Q1} = 1$.

$SD4 = PSD4 - (U + T)_{Q3} - (U + T)_{Q2} - U_{Q1} = 21 - 8 - 3 - 1 = 9$

$\Rightarrow Q_4 = SD4 / R_1 = 9 / 4 = 2$.

Example 4 – 10:

What is the quotient and remainder of 27483624 ÷ 6211?

D_1	D_2	D_3	D_4	D_5	D_6	D_7	D_8	
2	7	4	8	3	6	2	4	
R_1	R_2	R_3	R_4					
6	2	1	1					
Q_1	Q_2	Q_3	Q_4					
4	4	2	4					
PSD1	PSD2	PSD3	PSD4	PSD5	PSD6	PSD7	SD8	Remainder
	34	28	43					6160
SD1	SD2	SD3	SD4	SD5	SD6	SD7	SD8	
27	26	16	31					

Step 1

Compare D_1 and R_1

$D_1 = 2$ and $R_1 = 6$, thus $R_1 > D_1 \Rightarrow SD1 = D_1D_2 = 27$ and thus,

$\Rightarrow Q_1 = SD1 / R_1 = 27 / 6 = 4$.

Step 2

Instruction 1

$(UT + T)_{Q1} = R_1R_2 \times Q_1 = 62 \times 4 \Rightarrow 6 \times 4 = 24$ and $2 \times 4 = 08$

$\Rightarrow (UT + T)_{Q1} = 24 + 0 = 24$.

$PSD2 = [SD1 - (UT + T)_{Q1}] D_3 = [27 - 24]4 = 34$.

Mental Mathematics

Instruction 2

$(U + T)_{Q1} = R_2R_3 \times Q_1 = 21 \times 4 \Rightarrow 2 \times 4 = 08$ and $1 \times 4 = 04 = 08 + 0 = 8$.

$SD2 = [PSD2 - (U + T)_{Q1}] = 34 - 8 = 26 \Rightarrow Q_2 = SD2 / R_1 = 26 / 6 = 4$.

As there is no digit in the quotient before Q_1, Instruction 3, is not applicable here.

Step 3

Instruction 1

$(UT + T)_{Q2} = R_1R_2 \times Q_2 = 62 \times 4 \Rightarrow 6 \times 4 = 24$ and $2 \times 4 = 08 = 24 + 0 = 24$.

$PSD3 = [SD2 - (UT + T)_{Q2}] D_4 = [26 - 24]8 = 28$.

Instruction 2 and Instruction 3

$(U + T)_{Q2} = R_3R_4 \times Q_2 = 11 \times 4 \Rightarrow 1 \times 4 = 04$ and $1 \times 4 = 04 = 04 + 0 = 4$.

$(U + T)_{Q1} = R_2R_3 \times Q_1 = 21 \times 4 \Rightarrow 2 \times 4 = 08$ and $1 \times 4 = 04 = 8 + 0 = 8$.

$SD3 = [PSD3 - (U + T)_{Q2} - (U + T)_{Q1}] = [28 - 4 - 8] = 16$,

$\Rightarrow Q3 = SD3 / R_1 = 16 / 6 = 2$.

Step 4

Instruction 1

$(UT + T)_{Q3} = R_1R_2 \times Q_3 = 62 \times 2 = 12$ and $4 = 12 + 0 = 12$.

$PSD4 = [SD3 - (UT + T)_{Q3}] D_5 = [16 - 12]3 = 43$.

Instruction 2 and Instruction 3

$(U + T)_{Q3} = R_2R_3 \times Q_3 = 21 \times 2 \Rightarrow 2 \times 2 = 04$ and $1 \times 2 = 02 \Rightarrow (U + T)_{Q3} = 4 + 0 = 4$.

$(U + T)_{Q2} = R_3R_4 \times Q_2 = 11 \times 4 \Rightarrow 1 \times 4 = 04$ and $1 \times 4 = 04 \Rightarrow (U + T)_{Q2} = 04 + 0 = 4$.

$(U + T)_{Q1} = R_3R_4 \times Q_2$ (this is the same as $U_{Q1}) = R_4 \times Q_1 = 1 \times 4 = 4 = 4$.

$SD4 = [PSD4 - (U + T)_{Q3} - (U + T)_{Q2} - U_{Q1}] = 43 - 4 - 4 - 4 = 31$

Mental Mathematics

⇒ $Q_4 = SD4 / R_1 = 31 / 6 = 5$.

But $(UT + T)_{Q4} = R_1R_2 \times Q_4 = 62 \times 5$ ⇒ $6 \times 5 = 30$ and $2 \times 5 = 10$ ⇒ $(UT + T)_{Q4}$ $30 + 1 = 31$, which is not acceptable, because $(UT + T)$ should be less than SD.

Thus we reduce Q_4 by 1 ⇒ $Q_4 = (5 - 1) = 4$.

Remainder Determination

Now we will calculate the $(UT + T)_{Q4}$ for the real value of $Q_4 = 4$.

Instruction 1

$(UT + T)_{Q4} = R_1R_2 \times Q_4 = 62 \times 4 = 24$ and $08 = 24 + 0 = 24$.

$PSD5 = [SD4 - (UT + T)_{Q4}] D_6 = [31 - 24]6 = 76$.

Instruction 2 and Instruction 3

$(U + T)_{Q4} = R_2R_3 \times Q_4 = 21 \times 4$ ⇒ $2 \times 4 = 08$ and $1 \times 4 = 04$ $(U + T)_{Q4} = 08 + 0 = 8$.

$(U + T)_{Q3} = R_3R_4 \times Q_3 = 11 \times 2$ ⇒ $1 \times 2 = 02$ and $1 \times 2 = 02$ ⇒ $(U + T)_{Q3}$ $02 + 0 = 2$.

$U_{Q2} = R_4 \times Q_2 = 1 \times 4 = 4 = 04$.

$SD5 = [PSD5 - (U + T)_{Q4} - (U + T)_{Q3} - U_{Q2}] = 76 - 8 - 2 - 4 = 62$.

At this stage we will not calculate $(UT + T)$ anymore, but we will move D_7 to the right of SD5 to obtain the next pre–sub dividend, thus $PSD6 = SD5D_7 = 622$.

And then $(U + T)_{Q4} = R_3R_4 \times Q_4 = 11 \times 4$ ⇒ $1 \times 4 = 04$ and $1 \times 4 = 04$

⇒ $(U + T)_{Q4} = 4 + 0 = 4$.

And then $(U + T)_{Q3} = U_{Q3} = 1 \times 2 = 2$.

$SD6 = PSD6 - (U + T)_{Q4} - U_{Q3} = 622 - 4 - 2 = 616$.

⇒ $PSD7 = SD6D_8 = 6164$.

Now calculate $U_{Q3} = R_4 \times Q_4 = 1 \times 4 = 4$, and subtract it from PSD7 to obtain the remainder of the division operation:

Remainder = $[PSD7\ U_{Q3}] = 6164 - 4 = 6160$.

Mental Mathematics

Chapter 5

The Square of a Number

The multiplication of a number by itself is called the square of that number, for example, square of number 2, is 2×2 = 4, and the square of 6 is 6×6 = 36.

We use exponent sign to indicate the square of a number, for instance, the square of 2 is written as $2^2 = 4$, and a square of 6 as $6^2 = 36$.

We may use geometry to show the square of a number, and make the concept of square more clear, thus $2^2 = 4$ is represented, as a square with each side equal to 2 units, as shown in Table (5 – 1).

1	2
2	4

Table (5 – 1)

1	2	3	4	5	6	7	8	9
2	4							
3		9						
4			16					
5				25				
6					36			
7						49		
8							64	
9								81

Table (5 – 2)

In Table (5 – 2), diagonal of the square shows the square of digits 1 to 9.

The Square of a Two - Digit Number

One way of calculating the square of a two - digit number is to use the procedure already explained for multiplication of two - digit numbers, as in example 5 - 1.

If T_1U_1 and T_2U_2 are our two-digit numbers, we may use the following formula:

Mental Mathematics

Product = $[T_1 \times T_2][T1 \times U_2 + T_2 \times U_1][U_1 \times U_2]$, but $U_1 = U_2 = U$ and $T_1 = T_2 = T$.

⇨ Product = $[T^2][2T \times U][U^2]$

Begin from right to left, and whenever the result is a two digits number carryover the tenth digit to the left.

Example 5 – 1

What is $(23)^2 =?$

Solution:

1) $U = 3$, and $T = 2$ ⇨ $U^2 = 3^2 = 09$
2) $2 \times U \times T = 2 \times 2 \times 3 = 12$
3) T^2 ⇨ $2^2 = 04$
4) ⇨ $(23)^2 = [04][12][09]$ ⇨ $[4 + 1][2 + 0][9 = 529$.

Example 5 – 2

What is $(74)^2 =?$

Solution:

1) $4^2 = 16$,   2) $2 \times 7 \times 4 = 56$ and $56 + 1 = 57$,   3) $7^2 = 49$ and $49 + 5 = 54$,
2) ⇨ $(74)^2 = [49][56][16] = 5476$ ⇨ $[49][56 + 1][6]$
3) ⇨ $[49][57][6] = [49 + 5][7][6] = [49 + 5][7][6] = 5476$.

Trachtenberg Rapid Method, Special numbers, Type 1 and Type 2:

1) Special numbers Type 1 are two-digit numbers with digit 5, as their unit digits, such as 25, 75, and 95, and 95.and, in general, we may write them as N5, (N is 0 – 9).

In this case, 25 is the last right two digits of their square number and the left first digit (or first two digits), of the square number, is simply calculated by adding a 1 to the tenth digit and then multiply it by itself.

⇨ $(N5)^2 = [(N + 1) \times N][25]$

Mental Mathematics

For example, to obtain the square of 35, we add a 1 to 3, and multiply it by 3,

⇨ $(35)^2 = [(3 + 1) \times 3]25 = [4 \times 3]25 = 122$

Example 5 – 3

What is the square of 85?

Answer: $(N5)^2 = [(N + 1) \times N][25]$ ⇨ $(85)^2 = [(8 + 1) \times 8][25] = [9 \times 8][25] = 7225$.

1) Special numbers Type 2, are two-digit numbers, with 5 as their tenth digits, such as 53, 58, and 52, and in general, we may write them as 5N, (N is 0 – 9).

To obtain the square of Type 2, we may use the following formula:

$(5N)^2 = [25 + N][0N^2]$, if N ≤ 3, but if N > 3, then $(5N)^2 = [25 + N][N^2]$.

Note: the 5N square is always greater or equal to 2500, and always a four-digit number.

Example 5 – 4

What is the square of 52?

Answer –

As N ≤ 3 ⇨ $(52)^2 = [25 + 2][04] = 2704$.

Example 5 – 5

What is the square of 57?

Answer –

As N > 3 ⇨ $(57)^2 = [25 + 7][49] = 3249$.

General Procedure of Squaring 2-Digit Numbers

If MN is a two digit number, then we may use the following options to obtain the square of MN:

Mental Mathematics

Option A

1) Square the unit digit, $\Rightarrow N^2 = AB$
2) Multiply the tenth digit and unit digit, and multiply the result by 2, and then add it to the tenth digit of the number obtained from unit digit (in step 1)
   $\Rightarrow 2 \times M \times N = CD$ then $(CD + A) = EF$
3) Square the tenth digit, and add it to E.
4) $(MN)^2 = [M^2 + E][FB]$.

Option B

$MN = TU \Rightarrow (TU)^2 = [T^2][2 \times T \times U][U^2]$, this relation is already explained.

Example 5 – 6

What is $(73)^2 = ?$

1) $N^2 = 3^2 = 09 \Rightarrow A = 0$ and $B = 9 \Rightarrow AB = 09$.
2) $2 \times M \times N = 2 \times 7 \times 3 = 42 \Rightarrow (CD + A) = (42 + 0) = 42, \Rightarrow E = 4$, and $F = 2$.
3) $M^2 = 49$ thus $[M^2 + E] = [49 + 4] = 53$.
4) $(MN)^2 = [M^2 + E][FB] = 5329$.

Example 5 – 7

What is $(99)^2$?

1) $N^2 = 9^2 = 81 \Rightarrow A = 8$ and $B = 1$.
2) $2 \times M \times N = 2 \times 9 \times 9 = 162 \Rightarrow [CD + A] = [162 + 8] = 170, \Rightarrow E = 17$, and $F = 0$.
3) $M^2 = 81 \Rightarrow [M^2 + E] = [81 + 17] = 98$.
4) $(MN)^2 = [M^2 + E][FB] = 9801$.

Procedure of squaring Three-digit Numbers

1) **Special Cases, Type 1**

Type 1, three - digit numbers are of the form **MN5** (any three digit number, which ends to 5, such as 135, 725, and 285).

Mental Mathematics

To obtain the square of such numbers we use the following formula:

$(MN5)^2 = [(MN)^2 + MN][25]$

Example 5 – 8

What is the square of 115?

Answer –

$(MN5)^2 = [(MN)^2 + MN][25]$

$(115)^2 = [(11)^2 + 11][25] = [121 + 11][25] = 13225.$

Example 5 – 9

What is the square of 475?

$(475)^2 = [(47)^2 + 47]25 = [2256][25] = 225625.$

The Square of an Arbitrary Three Digit Number

In general, to obtain the square of an arbitrary three-digit number, we will do the following steps:

Step 1

Calculate the square of the last two digits from right to left.

Step 2

Multiply the first, and the last digit, and multiply the product by two.

Step 3

In this step, we add the first two digits of the square of the last two digits (we obtained in the first step), to the number we obtained in step – 2 (twice of the multiplication of the first and last digit).

Mental Mathematics

Step 4

Now we will calculate the square of the first two digits, but we will ignore the square of the middle digit, and then combine the rest with the number we obtained in step – 3, to obtain the square of our arbitrary number.

Now we will do some example to make the procedure as clear as possible.

Example 5 – 10

What is the square of 462?

Step 1

$(62)^2 = [6^2 = 36, \quad 2 \times 6 \times 2 = 24, \quad 2^2 = 04]$,

Now [do (U + T) addition of 36 and 24] and also, [do (U + T) addition of 24 and 04]

Thus for 36 and 24 we have, (U + T) = 6 + 2 = 8, and for 24 and 04 we have, 4 + 0 = 4.

$$\Rightarrow (62)^2 = 3844$$

Step 2

Multiply 4 and 2 and multiply the product by 2,

⇨ 4×2 = 8 and 2×8 = 16, and then add 16 to 38 of 3844 as follow: [16 + 38]44 = 5444.

Step 3

Now we will find the square of 46, the first two digits of N = 462 while ignoring the square of 6, and combine the rest with the square of 62, as follow:

$(46)^2$ = 16, 48, and 36, but we will ignore 6^2 = 36, and combine 16, and 48 with 5444.

Note: We ignored 36, square of 6, because the square of 6 is already used in the square of 62, and it should not be used twice.

Mental Mathematics

Step 4

⇨ $(462)^2$ = [(TU + T) of 20, and 13]3444 = [20 + 1]3444 = 213444.

1	6	(TU + T) of 16 and 48	4	8	(U + T) of 48 and 54	5	4	4	4
		16 + 4 = 20			8 + 5 = 13				
					(TU + U) of 20 and 13				
		Add 20 and 1 of 13			20 + 1 = 21	3			
					21	3	4	4	4
	$(462)^2$ =	213444							

How to calculate the square of a typical number?

We define a special or a typical number as a number consisting of the same digits, such as 111111, 222222, 33333333... And 99999999

There is a unique method of doing different arithmetic operations on them, which is very easy to perform and to remember.

In previous chapters, we introduced methods, of multiplication, addition, and division on them, and in this section, we will introduce very easy and rapid method for calculating their square.

Square of a number consisting of digit 1 only

If M is a number consist of N digit 1(N≤ 9), we use the following relation to obtain the square of M.

M^2 = ...[N – 3][N – 2][N - 1] N [N – 1][N – 2][N – 3] ....

Example 5 – 11

What is $(1111)^2$?

Answer –

N = 4, and M = 1111 ⇨ M² = [N – 3][N – 2][N - 1] N [N – 1][N – 2][N – 3]

⇨ $(1111)^2$ = 1234321

To obtain the square of such numbers, we don't need a pen or paper at all; but just understand the procedure and do it.

Mental Mathematics

Example 5 - 12

What is the square of (11111111)?

Answer: N = 8, ⇨ $(11111111)^2$ = 123456787654321.

Obviously, all digits of the square of such numbers are symmetric with respect to the number of digits which is placed in the middle of the square number.

Example 5 – 13

If every digit in number M is 1, and the number of digits in M is 9, how many digits are there in the square of M?

Answer –

1) 18

2) 17

3) 19

4) None of the above

Correct Answer is 17 why? Write a formula to support your answer.

Square of numbers consisting of digit 2 only
If M is a number consisting of the digit 2 only, such as 2222, 2222222, ...to obtain the square of M we do factorization of 2 on M, which will produce a number consisting of digit 1 multiplied by 2.

Now we will find the square of the number which consists of digit 1 and multiply the result by 2.

Example 5 – 14

What is the square of 22222?

Answer - 22222 = 2(11111), and thus $(22222)^2$ = $4(11111)^2$ = 4(123454321) = 493817284.

Square of numbers consisting of only digit 3, or 4, or...9
If a number consists of only digit 3 to 9, we will use the same method as for numbers consisting of only digit 2.

Mental Mathematics

Example 5 – 15

What is the square of 444444444?

Answer:

44444 = 4(11111) ⇨ (44444)^2 = (4)^2 × (11111)^2 = 16× (123454321) = 1975269136.

Square and Multiplication Table of one Digit Numbers

Here we will introduce a one-row table, which is used to do square and multiplication of one digit numbers, as follow:

10	9	8	7	6	6	7	8	9	10

In this table, we have 10 squares, two digits 6 in the middle of the table, and numbers 7, 8, 9, and 10 to the left and the right of them.

Example 5 – 16

What is the product of 7×8?

Solution – to find the tenth digit of the product choose 7 on one side, and 8 on the other side, and then count the number of the square between 7 and 8 including squares for 7 and 8, which is 5.

To find the unit digit, count number of squares from 7 to 10, and the number of squares from 8 to 10 (excluding 7, and 8), and multiply these two numbers, thus we 2×3 = 6,

⇨7×8 = 56

Example 5 – 17

What is the square of 8?

Answer -

the number of squares from 8 to 8 (including 8) is 6, and we have 2 squares to one side of 8 and 2 squares to another side of 8 (excluding 8), thus square of 8 is [6][2×2] = 64.

Chapter 6

The Square Root of a Number

As already explained in previous chapter, the square of a number is the product of the same number with itself, for example, the product of 11×11 is the square of the number 11.

The inverse operation of the square of a number is the square root of that number, for instance, the square of 25, is 25×25 = 625, but the square root of 25, is 5 because 5×5 = 25.

We use the exponent to represent the square of a number, thus $(25)^2$, is the same as 25×25.

To represent the square root of a number, we use the sign √, which is called radical, thus, the square root of 25, is represented as √25 = 5, which is the absolute value of the square root of 25, (in algebra √25 = ± 5).

Now that the notion of the square root is clear to us, we will explain how to determine the square root of a number, for special cases, and in general, using the Trachtenberg Method.

Note - We may use the table we introduced in chapter 5 to determine the square root of some, one digit numbers, and two digit numbers.

The root square of a one digit number is obviously a one digit number, and the square root of a two digit number is also a one digit number, because 99 is the biggest two digit number, and its square root is 9, with 18 as the remainder, and 10 is the square root of 100, which is the smallest three-digit number.

The root square of a Three - digit and a Four - digit number, lie between 100 to 9999, and thus the square root of a Three - digit number, and a Four - digit number is a two digit number, because √100 = 10, and √9999 = 99, with 198 as its remainder.

Determining the three - digit and four - digit, numbers square - root

Now with the help of few examples, we will explain the Trachtenberg procedure, to determine the 3 – digit, and 4 – digit numbers square root, as follow:

Mental Mathematics

Example 1 – 6

What is the square root of number 644?

Answer - :

Step 1

√644 = AB, where each of A and B is a one digit number.

Step 2

Divide 644 into two groups, with the help of an oblique line, as shown, 6 / 44.

Step 3

Now determine the two numbers, that are the perfect square, and 6 lies between them, and choose the square root of the smaller number as the square root of 6.

Step 4

Number 6 lies between two perfect square numbers of 4, and 9 (4<6<9).

Thus $2^2<6<3^2$, therefore square root of 6 is 2, which is the first digit of the square root of number 644(which means A = 2).

Step 5

Now subtract A^2 = 4, from 6 in number 644, as follow:

√644

-4

244.

Step 6

Now divide (6 – 4) by 2, and multiply it by ten, and then divide it by A, to obtain B, the second digit of the square root of 644, as below:

100

Mental Mathematics

(6 − 4)/ 2 = 1 and then B = (1×10)/A = (10/ 2) = 5, thus B = 5 ⇨ **AB = 25**.

Step 7

In this step, we will determine the remainder of the square root operation.

1) Determine the square of AB = 25.

$(25)^2$ = 4, 2×2×5, 25, or 4, 20, 25, and then ignore 4(we have already used this one), and combine 20 and 25 as below:

(20,   25) = 2[0 + 2]5 = 225, and now subtract 225 from 244, to obtain the remainder ⇨ **remainder = 244 − 255 = 19**.

Note - Obviously $(25)^2$ + 19 = 625 + 19 = 644

Example 6 − 2

What is √2200?

Answer −

Step 1

√2200 = AB, where each of A and B is a one digit number.

To determine A and B Divide number 2200 into two groups and countinue as follow:

22 / 00, but 16 < 22 < 25 thus 4^2 < 22 < 5^2 ⇨ A = 4.

Step 2

```
      √2200
  -    16
      _____
       0600
```

Step 3

(22 − 16)/ 2 = 3, now multiply it by 10, and divide it by A = 4 thus B = 30)/ 4 = 7.

Mental Mathematics

Step 4

Find square of AB, as follow:

$(AB)^2 = (47)^2 = 16$, 2×4×7, 49 or 16, 56, 49, now we ignore 16, and combine 56, and 49,

Thus we have 5[6 + 4]9 = 5[10]9 = (5 + 1)09 = 609.

But 609 > 600 thus we have to do correction, and reduce B = 7 by 1, and therefore B = 7 -1 = 6, ⇨ **AB = 46.**

Step 5

Now we will examine AB = 46 and determine the remainder of the square root operation.

$(AB)^2 = (46)^2 = 16$, 2×4×6, 36, or 16, 48, 36, now we will ignore 16, and combine 48, and 36.

4[8 + 3]6 = 4[11]6 = (4 +1)16 = 516, but 516 < 0600, thus 46 is the right answer.

Step 6

Subtract 516 from 0600, to obtain the remainder of square root operation.

⇨**remainder = 0600 – 516 = 84**.

Now we will examine the result as follow:

$(AB)^2 + 84 = (46)^2 + 84 = [16, 48, 36] + 84 = 16, 4[8 + 3]6 + 84 = 16, 4[11]6 + 84 = 16, [4 +1]16 = 16, 516 + 84 = 1[6 + 5]16 + 84 = 1[11]16 + 84 = 2116 + 84 = 2200$, thus 46, is the right answer, and 84 is the remainder of the square root operation.

Example 6 – 3

Determine the square root of 3025 and its remainder.

Answer –

Mental Mathematics

Step 1

√3025 = AB where each of A and B is a one digit number.

Divide 3025 into two groups of 30 and 25 with the help of an oblique line.

But we have 25<30<36, ⇨5^2<30<6^2 ⇨A = 5, which is the square root of 25.

Step 2

$$\begin{array}{r} \sqrt{3025} \\ - \quad 25 \\ \hline 0525 \end{array}$$

Step 3

[(05)/2] = (2 or 3), in the case of an odd number, we have two option, but we always choose the bigger number, thus for the case of number 5, we choose 3.

Step 4

B = [(3×10)/A] = [(30)/5] ⇨B = 6 and ⇨ AB = 56.

Note – the result for the last digit of the square root is never finalized until we examine the result.

Step 5

In this step we will examine the result as follow:

$(AB)^2$ = $(56)^2$ = 25, 2×5×6 (=60), and 36, now we ignore 25, and combine 60, and 36, as follow:

6[0 + 3]6 = 636, but 636 is greater than 0525, and thus we have to do correction, and reduce B by 1,

Therefore B = (6 – 1) = 5 ⇨**AB = 55**.

Mental Mathematics

Step 6

Now we have to check correctness of AB = 55, as follow:

$(AB)^2 = (55)^2 = 25$, $2 \times 5 \times 5 = 50$, 25, now we ignore the first 25, and combine 50, and 25 as follow:

5[0 + 2]5 = 525, thus B = 5, is the correct result ⇨**AB = 55**.

Remainder = 525 − 525 = 0.

Anticipating the Number of Digits in the Square Root of a Number
Now, we will explain, how to anticipate the number of digits in the square root of a number.

1) the number of digits in a square root of a number with even digits, like 6, 8, 12, and 2N, is, exactly half of the number of digits in that number,

⇨The number of digits in the square root of a number with 2N digits is 2N/2 = N.

2) If the number of digits in a number is odd, like 3, 5, 13, and (2N + 1), in order to find the number of digits in its square root, we add a 1, to that number and divide it by 2,

⇨ If number of digits in a number is (2N + 1), the number of digits in its square root is [(2N + 1) + 1]/2 = (2N + 2)/2 = (N + 1).

Example 6 – 4

Determine the number of digits, in the following numbers:

M = 569874, G = 8793256411927, and H = 158973210025.

Answer –

For M it is 6/2 = 3, for G it is (13 + 1)/2 = 7, and for H, it is 12/2 = 6.

Example 6 – 5

What is the square root of 207936?

Mental Mathematics

Answer –

Step 1

2N = 6 thus the number of digits in the square root of 207936, is 2N/ 2 = 6/2 = 3, now asume √207936 = ABC, where each of A, B, and C is a digit.

Step 2

Divide the number 207936 into three groups, with the help of oblique lines, as shown below:

$\boxed{20/\ 79/\ 36}$

Step 3

Begin from the first group, which is the number 20, and determine the two perfect square numbers, in such a way that 20, lies between them, as follow:

16 < 20 < 25 ⇨ 4^2 < 20 < 5^2 ⇨ A = 4.

Step 4

Now we will determine B, as follow:

20 − 16 = 4, and then [4/2] ×10 = 20 ⇨ B = [20/ A] = 20/4 = 5, and thus AB = 45.

Step 5

In this step we will determine C, the last digit of the square root of 207936, as follow:

$(AB)^2$ = $(45)^2$ = 16,   2×4×5 = 40,   25, and then ignore 16(it is already used), and combine 40, and 25, as shown here: 4[0 + 2]5 = 425.

Mental Mathematics

Step 6

$\sqrt{207936}$ (subtract $A^2 = 16$, from 20 and move 7 to its right)

-16

47

Now subtract 42 from 47 (the first two digits of 425, obtained in step 5).

47 − 42 = 5, and then divide it by 2.

[5/2] = 2 OR 3, and then multiply them by 10, and find their average, as follow:

{[(2×10) + (3×10)]/2} = 50/2 = 25, and then determine, C as follow:

$\boxed{C = [25/A] = [25/4] = 6.}$
$\sqrt{207936}$ = ABC, and ABC = 456

Step 7

Now we will do open consecutive multiplication, on ABC = 456, which is the multiplication of 4, and 6, and twice the result ⇨ 2×A×C = 2×4×6 = 48.

Step 8

Determine the square of BC = 56.

$(56)^2$ = 25,   2×5×6 = 60,   36, and ignore 25, and then combine 60, and 36, as follow:
6[0 + 3]6 = 636.

Step 9

Now we will review, our operation on $\sqrt{207936}$, as it is shown below, first, we subtracted 16, and then 42 (from 425, we obtained in step 5)

Mental Mathematics

√207936

- 16

  047936
  -42

  005936
  And now we will subtract 4 (of 48 that we obtained in step 7)
  005936
  004

  001936.

Step 10

Now we will update and combine 425, 48, and 636.

We already used 42 of 425, thus we are left with 5, and then we used 4 from 48, and thus here we are left with 8.

Now we will add 5, 8, and 636, as follow:

5

8

636

1936 = (001936).

Now if we subtract 1936, from 001936(obtained in step 9), we will obtain the remainder of the square root operation, 001936 – 001936 = 0

Which means ABC = 456, is the perfect square root of 207936, with zero remainder.

Mental Mathematics

Example 6 – 6

What is √364728?

Answer –

Step 1

Divide number 364728 into groups of two digits, using an oblique line to separate them from each other, as shown here: 36 / 47 / 28.

Step 2

As we have 6 digits in 364728, therefore, the number of digits in the square root of number 364728 is 2N/2 = 6/2 = 3. And now we will assume that ABC to be the square root of number 364728, where each of A, B, and C is just a digit.

Step 3

We will determine A, which is the square root of 36, as follow:

36 ≤ 36 < 49, ⇨ $6^2 ≤ 36 < 7^2$ ⇨ A = 6.

Step 4

Now we will determine B as follow:

√36 47 28

  36

  00, and then (00/2) = 0, then multiply 0, by 10, and divide it by A to find B,

⇨ B = [(0)10/A] = (0)/6 = 0.

Step 4

In this step we will determine C, the last digit of the square root, as follow:

Mental Mathematics

√36 47 28

   36

_____

00 04

Now $(60)^2 = 36$, $2(6) \times (0)$, $(0)^2 = 36$, 00, $(0)^2 = 00$, but we ignore 36, and by combining 00, and 00, we obtain 000, and then subtract first 0, from $36 - 36 = 00$, resulting in 0, with 0 reminder, which goes to the left of 4.

04

00

04/ 2 = 2 and then C = (2×10)/A = 20/6 = 3.

Step 5

Therefore, ABC = 603, and now we will determine the remainder of the square root operation.

Now subtract the second 0 of 000(36 -36 = 000) from 04(04 – 0 = 04).

Thus we have:

√36 47 28

   36

_____

00 04

Step 6

Now multiply A, and C, and multiply the result by 2:

2×6×3 = 36, and subtract 3 of 36 from 04(4 – 3 = 1), and move 1 to the left of 7:

00 04 17, and then move 2, and 8 also to the right of 00 04 17, thus we obtain:

Mental Mathematics

00 04 17 28

Step 7

Now determine the square of BC = 03, thus we have $(03)^2$ = 00, 2(0) × (3), 09

= 00, 09.

Now add up 000, 36, and 09 as follow:

~~000~~
  ~~36~~
    009

    609

(00 and 3 are already used, so we don't use them again).

Step 8

Now we subtract 609, from 1728, to determine the remainder of the square root operation.

1728 (these numbers were not used while doing operation on 364728)

  609 (these numbers were not used while doing operation on ABC = 603)

  1119 Remainder of the square root operation.

Obviously, we have $(603)^2 + 1119 = 364728$.

Example 6 – 7

What is $\sqrt{893304}$?

Answer:

Mental Mathematics

Step 1

√893304 = ABC because there are only 6 digits in the number 893304, thus, its square root, has 3 digits only.

Divide number 89334 into three groups with the help of oblique lines as follows:

89/33/04, and now we will proceed the same way as in previous examples in this section ⇨81 < 89 < 100 ⇨9^2<89<10^2 ⇨ A = 9.

Step 2

89 – 81 = 8, and then divide 8 by 2 ⇨ 8/2 = 4 ⇨B = (4×10)/A = 40/9 = 4.

Step 3

$(AB)^2$ = $(94)^2$ = 81, 2×9×4, 16 =81, 72, 16, now we ignore 81(it is already used), and combine 72, and 16.

72,   16 = 7[2 +1]6 = 736

Step 4

In this step we will determine C, the third digit of the square root, as follow:

√8 9 3 3 0 4 -

  81

  08

   7

   1

Subtract 7(first digit of 736 obtained in step3), from 08, and move 3, the third digit of 893304 to the right of 1).

08 13, and now subtract 3, (the second digit of 736), from 13.

Now as the result of these two consecutive subtractions we have 08, and 10.

Mental Mathematics

Now divide 10, by 2, and multiply the result by 10, and divide it by A = 9, as below:

10/2 = 5 ⇨ C = [(5×10)/A)] = 50/9 = 5 ⇨ ABC = 945.

Step 5

In this step we will find the remainder of the square root operation, as follow:

$(BC)^2 = (45)^2$ = 16, 2×4×5, 25 =16, 40, 25, now will ignore 16, and combine 40, and 25, as follow:

40, 25 = 4[0 + 2]5 = 425

Now multiply 9, and 5(the first and the last digit of ABC = 945), and multiply the result by 2 and divide the result by 10.

2×9×5/10 = 9 and subtract 9 from 10 ⇨10 – 9 = 1, and then move 1 to the right of 3, the fourth digit of the number 893304, and also, move it's 04, to the right of 81, 13, 13, and obviously we will obtain the following result out of operations, don on number 893304.

81 13 1304, but we have already used 81, and first 13, therefore the final number left is 1304.

Step 6

In this step we will determine the final result of all operations done on the square root (ABC = 945), as below:

7 3 6

  90

    425

[4 + 6]25 = 1025

Step 7

Mental Mathematics

Now subtract 1025 from 1304, to determine the remainder of the square root operation, 1304 – 1025 = 279.

⇨**Square root of 893304 is, 945, with the remainder of 279.**

Chapter 7

Proof of Correctness

The aim of this chapter is not to present the Proof of the correctness of every method explained in previous chapters, but to open a window for readers, how these ideas could be proved.

We will prove the correctness of very few procedures explained in previous chapters and leave the rest as a research and practice for readers.

Readers of this book are a wide range of people with a different mathematical background, therefore we may describe something here which is known to some of you, but the same thing could be new and interesting for other readers.

However, if you are already familiar with these definitions and idea, you may escape it, and proceed to the following sections.

We will mostly use Algebra, in our proofs, but may use basic number theory, geometry, or other topics of mathematics as well.

Definition 7 – 1
The set of natural numbers is the set of the positive whole numbers, as follow:

1, 2, 3, 4, 5, 6, 7, 8, 9, 10……………,

Definition 7 – 2
The decimal number system, is a system of numbers, that uses digits, 0 – 9 as its alphabets, and each number is expressed in base ten, therefore the value of each digit in a number in this system depends on its position in that number.

If M is a number in the decimal system, its left digit is the most valuable digit, which is called the most significant digit, and its right digit is the least valuable digit, which is called the least significant digit.

For example, if M = 742913, then 7 is the most significant digit, and 3 is the least significant digit.

Mental Mathematics

Formula 7 – 1

Now if we assume N as a number in the decimal system with m digits, then we have:

$N = N_{(m-1)} N_{(m-2)} \ldots\ldots N_{(m-(m-1))} N_{(m-m)}$, where $N_{(m-1)}$ is the most significant digit, and $N_{(m-m)}$, is the least significant.

$\Rightarrow N = [N_{(m-1)}] \times 10^{(m-1)} + [N_{(m-2)}] \times 10^{(m-2)} + \ldots\ldots + [N_{(m-(m-1))}] \times 10^{(m-(m-1))} + [N_{(m-m)}] \times 10^{(m-m)}$

Therefore for N = 742913 we have m = 6 and thus:

$N = 7 \times 10^{(6-1)} + 4 \times 10^{(6-2)} + 2 \times 10^{(6-3)} + 9 \times 10^{(6-4)} + 1 \times 10^{(6-5)} + 3 \times 10^{(6-6)}$

$\Rightarrow N = 700000 + 40000 + 2000 + 900 + 10 + 3.$

The Correctness Proof of Number N by 11 rules

Derive a multiplication formula for number N by 11, with m digits in N.

As we explained in chapter 1, the general procedure is to add, each digit of N, to its right digit, and if in any step the outcome is a two digit number, carry over the left digit of the outcome to the next left digit, and do the same thing with every digit in N.

If N is a number with m digit, the following relations proves that the above procedure is correct for any number.

1) $11 = (10 + 1) = 1 \times 10^1 + 1 \times 10^0$
2) $N = N_{(m-1)} \times 10^{(m-1)} + N_{(m-2)} \times 10^{(m-2)} + \ldots\ldots + N_2 \times 10^2 + N_1 \times 10^1 + N_0 \times 10^0$
3) $N \times 11 = [N_{(m-1)} \times 10^{(m-1)} + N_{(m-2)} \times 10^{(m-2)} + \ldots\ldots + N_2 \times 10^2 + N_1 \times 10^1 + N_0 \times 10^0] \times [1 \times 10^1 + 1 \times 10^0]$
4) $\Rightarrow N \times 11 = (10) \times [[N_{(m-1)} \times 10^{(m-1)} + N_{(m-2)} \times 10^{(m-2)} + \ldots\ldots + N_2 \times 10^2 + N_1 \times 10^1 + N_0 \times 10^0] + (1) \times [N_{(m-1)} \times 10^{(m-1)} + N_{(m-2)} \times 10^{(m-2)} + \ldots\ldots + N_2 \times 10^2 + N_1 \times 10^1 + N_0 \times 10^0]$

$\Rightarrow N \times 11 = [N_{(m-1)} \times 10^m + N_{(m-2)} \times 10^{(m-1)} + \ldots\ldots + N_2 \times 10^3 + N_1 \times 10^2 + N_0 \times 10^1]$

$+ [N_{(m-1)} \times 10^{(m-1)} + N_{(m-2)} \times 10^{(m-2)} + \ldots\ldots + N_2 \times 10^2 + N_1 \times 10^1 + N_0 \times 10^0]$

Mental Mathematics

Now we will do factorization of the terms with a factor 10 and the same exponent.

$\Rightarrow N \times 11 = \{N_{(m-1)} \times 10^m + [N_{(m-1)} + N_{(m-2)}] \times 10^{(m-1)} + \ldots + [N_1 + N_2] \times 10^2 + [N_0 + N_1] \times 10^1 + N_0 \times 10^0\}$

Now we write the above relation in the decimal number system format form as follow:

$\Rightarrow N \times 11 = [N_{(m-1)}] [N_{(m-1)} + N_{(m-2)}] \ldots [N_1 + N_2] [N_0 + N_1][N_0]$

Where $[N_{(m-1)}]$ is the most significand digit and $[N_0]$ is the least significat digit of number N×11.

Example 7 – 1

What is the product of 295×11 =?

Answer – Now we will follow the same procedure explained in driving multiplication formula for number N by 11.

$\Rightarrow 295 = 2 \times 10^2 + 9 \times 10^1 + 5 \times 10^0$

$\Rightarrow 295 \times 11 = [2 \times 10^2 + 9 \times 10^1 + 5 \times 10^0] \times [10 + 1] = 2 \times 10^3 + 9 \times 10^2 + 5 \times 10^1 + 2 \times 10^2 + 9 \times 10^1 + 5 \times 10^0$

Now we will do factorization of the terms with a factor 10 and the same exponent.

$\Rightarrow 295 \times 11 = 2 \times 10^3 + [2 + 9] \times 10^2 + [9 + 5] \times 10^1 + 5 \times 10^0$

Now we write the above relation in a decimal system number format form as below:

$\Rightarrow 295 \times 11 = [2 + C][2 + 9 + C][9 + 5]5 = [2 + C][11 + C][14]5$

Note – Each C in a square bracket is the carryover from its right digit.

$\Rightarrow 295 \times 11 = [2 + C][12][4]5 = [2 + 1]245 = 3245$.

Mental Mathematics
The Correctness Proof of Number N by 6 rules

Derive a multiplication formula for number N by 6, with m digits in N.
Answer –

For the ease of understanding, we choose N as a four digit number,

⇨ N = ABCD, where each of A, B, C, and D is a single even digit number.

Now we use the same logic and procedure for N = ABCD as explained earlier, for number N by 11 as follow:

Step 1

$N = ABCD \Rightarrow N = A \times 10^3 + B \times 10^2 + C \times 10^1 + D \times 10^0$, but $6 = (1/2) \times 10 + 1$.

⇨ $6 \times N = [A \times 10^3 + B \times 10^2 + C \times 10^1 + D \times 10^0] \times [(1/2) \times 10 + 1]$

If we put a zero on the left-hand side of ABCD, then we have N = 0ABCD, and we assume a zero on the right side of D also (but this zero is not counted as a digit of number N).

Step 2

⇨ $6 \times N = [0 + (1/2) A] \times 10^4 + [(1/2) B] \times 10^3 + [(1/2) C] \times 10^2 + [(1/2) D] \times 10^1 + A \times 10^3 + B \times 10^2 + C \times 10^1 + D \times 10^0$

Now we will do factorization of the terms with a factor 10 and the same exponent.

⇨ $6 \times N = [0 + (1/2) A] \times 10^4 + [A + (1/2) B] \times 10^3 + [B + (1/2) C] \times 10^2 + [C + (1/2) D] \times 10^1 + [D + (1/2)0] \times 10^0$

Now we write the above relation in the decimal system number format form as below:

⇨ $6 \times N = [0 + (1/2) A][A + (1/2) B][B + (1/2) C][C + (1/2) D][D + (1/2)0]$

Now we will do an example, assuming that all digits in N are even (such as 0, 2, 4. 6, and 8).

Example 7 – 2

What is the product of number N = 6482, by 6?

Mental Mathematics
Answer –

To obtain the product of 6×6482, we will do the following steps:

Step 1

$6482 = 6 \times 10^3 + 4 \times 10^2 + 8 \times 10^1 + 2 \times 10^0$, and $6 = (1/2) \times 10 + 1$.

Step 2

⇨ $6 \times (6482) = [(1/2) \times 10 + 1] \times [6 \times 10^3 + 4 \times 10^2 + 8 \times 10^1 + 2 \times 10^0]$

⇨ $6 \times (6482) = [(1/2) \times 6 \times 10^4 + (1/2) \times 4 \times 10^3 + (1/2) \times 8 \times 10^2 + (1/2) \times 2 \times 10^1] + [6 \times 10^3 + 4 \times 10^2 + 8 \times 10^1 + 2 \times 10^0]$

Now we will remove the square brackets, and do factorization of the terms with a factor 10 and the same exponent.

⇨ $6 \times (6482) = [0 + (1/2) \times 6] \times 10^4 + [6 + (1/2) \times 4] \times 10^3 + [4 + (1/2) \times 8] \times 10^2 + [8 + (1/2) \times 2] \times 10^1 + [2 + (1/2) \times 0] \times 10^0$

$= 3 \times 10^4 + 8 \times 10^3 + 8 \times 10^2 + 9 \times 10^1 + 2 \times 10^0 = 38892$.

If a digit in number N is odd, such as 1, 3, 5, 7, and 9, then we have to modify the formula as follow:

Again we consider N = ABCD, which is a four digit number, assuming B to be odd,

⇨ $B = 2m + 1$, and obviously, m is the smaller half of B.

Now we put a zero to the left of ABCD.

⇨ N = 0ABCD ⇨ N = 0A [2m + 1] CD

⇨ $N = 0 \times 10^4 + A \times 10^3 + (2m + 1) \times 10^2 + C \times 10^1 + D \times 10^0$ and $6 = (1/2) \times 10^1 + 1 \times 10^0$

⇨ $6 \times N = [(1/2) \times 10 + 1] \times [0 \times 10^4 + A \times 10^3 + (2m + 1) \times 10^2 + C \times 10^1 + D \times 10^0]$

⇨ $6 \times N = [0 \times 10^5 + (1/2) \times A \times 10^4 + (1/2) \times (2m + 1) \times 10^3 + (1/2) \times C \times 10^2 + (1/2) \times D \times 10^1] + [0 \times 10^4 + A \times 10^3 + (2m + 1) \times 10^2 + C \times 10^1 + D \times 10^0]$

Now we will do factorization of the terms with a factor 10 and the same exponent.

Mental Mathematics

⇨ 6×N = [0 + (1/2) ×A] ×10⁴ + [A + (1/2) × (2m + 1)] ×10³ + [(2m + 1) + (1/2) × C] ×10² + [C + (1/2) × D] ×10¹ + [D + (1/2) ×0] ×10⁰

⇨ 6×N = [0 + (1/2) ×A] ×10⁴ + [A + m + 1/2] ×10³ + [B + (1/2) × C] ×10² + [C + (1/2) × D] ×10¹ + + [D + (1/2) ×0] ×10⁰

But, [A + m + 1/2] ×10³ = (A + m) ×10³ + 5×10²

⇨ 6×N = [0 + (1/2) ×A] ×10⁴ + (A + m) ×10³ + [B + (1/2) × C + 5] ×10² + [C + (1/2) × D] ×10¹ + + [D + (1/2) ×0] ×10⁰

= [(1/2) ×A][A + m][B + (1/2) × C + 5][C + (1/2) × D] D.

Note – In the above relation we have replaced B for 2m + 1, where it is combined with one-half of C, but where it is combined with A it is left as m, which is the smaller half of the B.

The above relation proofs the correctness multiplication rule of 6 by a number, as follow:

Add each digit to the half of its right digit, and if a digit is odd, add a five to it as well.

Example 7 – 3

What is the product of 8726 by 6?

Ansewr:

[(1/2) ×A][A + m][B + (1/2) × C + 5][C + (1/2) × D] D. = [0 + 8/2 +carryover] [8 + 3 + carryover] [7 + 2/2 + 5][2 + 6/2][6 + 0/2]

=52356.

As it is shown in the above example, we start our operation from right, and moved to the left, and whenever result is a two-digit number we carry over a 1, to the left digit.

The Correctness Proof of Number N by 5 rules
Derive a multiplication formula for number N by 5, with m digits in N.
Answer:

Mental Mathematics

For the sake of easiness, we assume N to be just a four digit number, but we may do the same thing for any bigger number.

Step 1

N = ABCD, and 5 = (1/2) ×10

Step 2

We put a zero on the left side of A, and a zero on the right of D (Zero on the right side of D is not considered a digit of number N).

⇨ N = [0ABCD] = 0×10^4 + A×10^3 + B×10^2 + C×10^1 + D×10^0 + 0×10^0

⇨ N×5 = [(1/2)10] × [0×10^4 + A×10^3 + B×10^2 + C×10^1 + D×10^0 + 0×10^0]

⇨ N×5 = [(1/2)0 ×10^5 + (1/2) A ×10^4 + (1/2) B×10^3 + (1/2) C×10^2 + (1/2) D×10^1 + (1/2) 0×10^0]

Now we write the above relation in the decimal number system format form as below:

⇨ 5×N =[(1/2)0][(1/2)A][(1/2)B][(1/2)C][(1/2)D][(1/2)0]

Above relation works if every digit in N is even.

Example 7 – 4

If N = 8266402, what is 5×N?

Answer:

N = 0[8266402]0

5×N = [(1/2)0][(1/2)8][(1/2)2][(1/2)6][(1/2)6][(1/2)4][(1/2)0][(1/2)2][(1/2)0]

= 041332010.

Now we will assume one of the digits in N, to be an odd digit, like 1, 3, 5.7, or 9.

Again we assume N = 0[ABCD] 0, and also B = 2m + 1.

Mental Mathematics

$\Rightarrow 5 \times N = [(1/2)0 \times 10^5 + (1/2) A \times 10^4 + (1/2)(2m + 1) \times 10^3 + (1/2) C \times 10^2 + (1/2) D \times 10^1 + (1/2) 0 \times 10^0]$

$= [(1/2)0 \times 10^5 + (1/2) A \times 10^4 + (m + 1/2) \times 10^3 + (1/2) C \times 10^2 + (1/2) D \times 10^1 + (1/2) 0 \times 10^0]$

$= [(1/2)0 \times 10^5 + (1/2) A \times 10^4 + (m) \times 10^3 + 5 \times 10^2 + (1/2) C \times 10^2 + (1/2) D \times 10^1 + (1/2) 0 \times 10^0]$

$= [(1/2)0 \times 10^5 + (1/2) A \times 10^4 + (m) \times 10^3 + [5 + (1/2) C] \times 10^2 + (1/2) D \times 10^1 + (1/2) 0 \times 10^0]$

$= [0][(1/2) A][m][5 + (1/2)C][(1/2)D][0]$

Example 7 – 5

If N = 6582, then what is 5×N?

Answer:

$5 \times N = [0][(1/2) A][m][5 + (1/2)C][(1/2)D][0]$

$= 032[5 + 4]10 = 032910$

Note: m = 5/2 = 2, which is the smaller half of 5.

The Correctness Proof of Number N by 9 rules
Derive a multiplication formula for number 9 by N with m digits long.

For the purpose of easiness, we assume number N with four digits, as A, B, C, and D, but we may use the same procedure if N has more than four digits.

Step 1

N = ABCD, and 9 = 10 – 1.

$\Rightarrow ABCD = A \times 10^3 + B \times 10^2 + C \times 10^1 + D \times 10^0$

$\Rightarrow 9 \times N = (10 - 1) \times [A \times 10^3 + B \times 10^2 + C \times 10^1 + D \times 10^0]$

$\Rightarrow 9 \times N = A \times 10^4 + B \times 10^3 + C \times 10^2 + D \times 10^1 - A \times 10^3 - B \times 10^2 - C \times 10^1 - D \times 10^0$

Mental Mathematics

As you know if we add a number to another number, and then subtract the same number, the value of that number remains the same.

For example $5 = (5 + 3 - 3)$, and $M = M + N - N$.

Now we will add, and subtract $(9 \times 10^3 + 9 \times 10^2 + 9 \times 10^1 + 9 \times 10^0)$ to $9 \times N$.

$\Rightarrow 9 \times N = A \times 10^4 + B \times 10^3 + C \times 10^2 + D \times 10^1 - A \times 10^3 - B \times 10^2 - C \times 10^1 - D \times 10^0 + (9 \times 10^3 + 9 \times 10^2 + 9 \times 10^1 + 9 \times 10^0) - (9 \times 10^3 + 9 \times 10^2 + 9 \times 10^1 + 9 \times 10^0)$.

Now we will do partial factorization of common foctors among chosen terms as follow:

$\Rightarrow 9 \times N = A \times 10^4 + [9 - A + B] \times 10^3 + [9 - B + C] \times 10^2 + [9 - C + D]\, 10^1 + [9 - D]\, 10^0$

$\quad - (9 \times 10^3 + 9 \times 10^2 + 9 \times 10^1 + 9 \times 10^0)$

But, $(9 \times 10^3 + 9 \times 10^2 + 9 \times 10^1 + 9 \times 10^0) = 9999$, and $9999 = 10000 - 1 = 10^4 - 1$ and $-9999 = -10^4 + 1$

$\Rightarrow 9 \times N = A \times 10^4 + [9 - A + B] \times 10^3 + [9 - B + C] \times 10^2 + [9 - C + D]\, 10^1 + [9 - D] \times 10^0 - 10^4 + 1 \times 10^0$.

Now we will do factrorization of 10^4 for $[A \times 10^4]$ and 10^4, and factrorization of 10^0 between 1×10^0 and $[9 - D] \times 10^0$.

$\Rightarrow 9 \times N = [A - 1] \times 10^4 + [9 - A + B] \times 10^3 + [9 - B + C] \times 10^2 + [9 - C + D]\, 10^1 + [9 + 1 - D] \times 10^0$

Now we write the above relation in a decimal system number format form as below:

$\Rightarrow 9 \times N = [A - 1][9 - A + B][9 - B + C][9 - C + D][10 - D]$

Please note each of $[A - 1]$, $[9 - A + B]$, $[9 - B + C]$, $[9 - C + D]$, $[9 - C + D]$, $[10 - D]$, is a digit of the product of $9 \times N$, which profs the same rule, we already explained in chapter 1.

Example 7 – 6

What is the product of 9×6583?

Mental Mathematics

Ansewr:

A = 6, B = 5, C = 8 and D = 3, thus we have:

9×6583 = [A − 1][9 − A + B][9 − B + C][9 − C + D][9 − C + D][10 − D]

= [6 − 1][9 − 6 + 5][9 − 5 + 8][9 − 8 + 3][10 − 3] = 59247.

Please note, it is always required to start our operation from right to the left, as we may encounter with carryover, which should be added to the next left digit.

The Correctness Proof of Number N by 8 rules
Derive a formula to find the product of number 8 by N with m digits long.

For the purpose of easiness, we assume number N with four digits, as A, B, C, and D, but we may use the same procedure for N if the number of digits in N is more than four.

N = ABCD, and 8 = 10 − 2.

⇨ $ABCD = A \times 10^3 + B \times 10^2 + C \times 10^1 + D \times 10^0$

⇨ $8 \times N = (10 - 2) \times (A \times 10^3 + B \times 10^2 + C \times 10^1 + D \times 10^0)$

$= A \times 10^4 + B \times 10^3 + C \times 10^2 + D \times 10^1 - 2A \times 10^3 - 2B \times 10^2 - 2C \times 10^1 - 2D \times 10^0$

Now we will add, and subtract $(18 \times 10^3 + 18 \times 10^2 + 18 \times 10^1 + 18 \times 10^0)$ to 8×N.

Thus we have:

⇨ $8 \times N = A \times 10^4 + B \times 10^3 + C \times 10^2 + D \times 10^1 - 2A \times 10^3 - 2B \times 10^2 - 2C \times 10^1 - 2D \times 10^0 +$

$18 \times 10^3 + 18 \times 10^2 + 18 \times 10^1 + 18 \times 10^0 - 18 \times 10^3 - 18 \times 10^2 - 18 \times 10^1 - 18 \times 10^0$

Now we will do factorization of the terms, with a factor 10, and the same exponent, as follows:

⇨ $8 \times N = A \times 10^4 + [18 - 2A + B] \times 10^3 + [18 - 2B + C] \times 10^2 + [18 - 2C + D] \times 10^1 +$

$[18 - 2D] \times 10^0 - 18 \times 10^3 - 18 \times 10^2 - 18 \times 10^1 - 18 \times 10^0$

But, $- 18 \times 10^3 - 18 \times 10^2 - 18 \times 10^1 - 18 \times 10^0 = - 19998$, and − 19998 = − 20000 + 2

123

Mental Mathematics

And $- 20000 + 2 = -2 \times 10^4 + 2$

$\Rightarrow 8 \times N = A \times 10^4 + [18 - 2A + B] \times 10^3 + [18 - 2B + C] \times 10^2 + [18 - 2C + D] \times 10^1 + [18 - 2D] \times 10^0 - 2 \times 10^4 + 2 \times 10^0$

Now we will do factorization of the items, $A \times 10^4$, and -2×10^4, as one group, and also $[18 - 2D] \times 10^0$, with 2×10^0, as another group, as shown below:

$8 \times N = [A - 2] \times 10^4 + [18 - 2A + B] \times 10^3 + [18 - 2B + C] \times 10^2 + [18 - 2C + D] \times 10^1 + [18 - 2D + 2] \times 10^0$

$\Rightarrow 8 \times N = [A - 2][18 - 2A + B][18 - 2B + C][18 - 2C + D][20 - 2D]$

Note – please note each of square bracket, in above relation is just a digit of $8 \times N$, product.

Example 7 – 7

What is the product of 8×7289?

Ansewr:

ABCD = 7289, and N = 8

$8 \times N = [A - 2][18 - 2A + B][18 - 2B + C][18 - 2C + D][20 - 2D]$

$8 \times N = [7 - 2][18 - 2 \times 7 + 2][18 - 2 \times 2 + 8][18 - 2 \times 8 + 9][20 - 2 \times 9]$

$\Rightarrow 8 \times 7289 = [5][8][3][1][2] = 58312.$

Please note, it is always required to start our operation from right to the left, as we may encounter with carryover, which should be added to the next left digit.

Multiplication Operation - Units Digit and Tens Digit Method

Assume that we have a three digit number, such as ABC, and want to multiply it by N, which is a digit also.

$\Rightarrow N \times (ABC) = [A \times 10^2 + B \times 10^1 + C \times 10^0] \times N$

Mental Mathematics

$\Rightarrow N \times (ABC) = [N \times A] \times 10^2 + [N \times B] \times 10^1 + [N \times C] \times 10^0$

As N is a digit (0 - 9), and each of A, B, or C is also a digit, thus each of the N×A, N×B, and N×C is also a number with two digits at most.

Note: Each of the N×A, N×B, and N×C lies between zero to eighty-one.

Now by using symbols, **U**, and **T** as **Units Digit** and **Tens Digit**, we may write, each of the N×A, N×B, and N×C, as follow:

$N \times A = [T_A] \times 10^1 + [U_A] \times 10^0$

$N \times B = [T_B] \times 10^1 + [U_B] \times 10^0$

$N \times C = [T_C] \times 10^1 + [U_C] \times 10^0$

Now by replacing, $[T_A] \times 10^1 + [U_A] \times 10^0$, for N×A, and $[T_B] \times 10^1 + [U_B] \times 10^0$, for N×B, and $[T_C] \times 10^1 + [U_C] \times 10^0$, for N×C, in $[N \times A] \times 10^2 + [N \times B] \times 10^1 + [N \times C] \times 10^0$, we obtain:

$[N \times A] \times 10^2 + [N \times B] \times 10^1 + [N \times C] \times 10^0 =$

$[(T_A) \times 10^1 + (U_A) \times 10^0] \times 10^2 + [(T_B) \times 10^1 + (U_B) \times 10^0] \times 10^1 + [(T_C) \times 10^1 + (U_C) \times 10^0] \times 10^0$

$\Rightarrow N \times (ABC) = (T_A) \times 10^3 + (U_A) \times 10^2 + (T_B) \times 10^2 + (U_B) \times 10^1 + (T_C) \times 10^1 + (U_C) \times 10^0$

Now we will do factorization of the terms, with the same factor:

$\Rightarrow N \times (ABC) = (T_A) \times 10^3 + [(U_A) + (T_B)] \times 10^2 + [(U_B) + (T_C)] \times 10^1 + (U_C) \times 10^0$

$\Rightarrow \mathbf{N \times (ABC) = [T_A][(U_A) + (T_B)] [(U_B) + (T_C)][U_C]}$

These formulas prove the correctness of the idea, we explained in chapter 2.

Example 7 – 8

What is the product of 287×4?

Mental Mathematics

Answer:

⇨ ABC = 287, and N = 4 ⇨ A = 2, B = 8, and C = 7.

4×7 = 28 thus U_C = **8**, and **Tc = 2**

4×8 = 32, thus U_B = **2**, and T_B = **3**.

4×2 = 08, thus U_A = **8**, and T_A = **0**.

⇨ $[T_A][(U_A) + (T_B)] [(U_B) + (T_C)][U_C]$ = 0[8 + 3][2 + 2][8] = 0[11]58 = [0 + 1]158 = 1148.

Multi Digit Numbers Multiplication – Units Digit and Tens Digit Method

Now we will multiply a three digit number ABC, by a two digit number MN.

$(ABC) \times (MN) = [(A \times 10^2 + B \times 10^1 + C \times 10^0)] \times [M \times 10^1 + N \times 10^0]$

$= (M \times A) \times 10^3 + (M \times B) \times 10^2 + (M \times C) \times 10^1 + (N \times A) \times 10^2 + (N \times B) \times 10^1 + (N \times C) \times 10^0$

As each of N, and M is a digit, and each of A, B, or C is also a digit, thus each of the N×A, N×B, N×C, M×A, M× B, and M×C is a two digit number too (0 – 81).

Now by using symbols, **U**, and **T** as **Units Digit** and **Tens Digit**, we may write, each of the N×A, N×B, N×C, M×A, M× B, and M×C, as follow:

N×A = $[T_{NA}] \times 10^1 + [U_{NA}] \times 10^0$

N×B = $[T_{NB}] \times 10^1 + [U_{NB}] \times 10^0$

N×C = $[T_{NC}] \times 10^1 + [U_{NC}] \times 10^0$

M×A = $[T_{MA}] \times 10^1 + [U_{MA}] \times 10^0$

M×B = $[T_{MB}] \times 10^1 + [U_{MB}] \times 10^0$

M×C = $[T_{MC}] \times 10^1 + [U_{MC}] \times 10^0$

Now by plugging, $[T_{NA}] \times 10^1 + [U_{NA}] \times 10^0$, for N×A, and $[T_{NB}] \times 10^1 + [U_{NB}] \times 10^0$, for N×B, and $[T_{NC}] \times 10^1 + [U_{NC}] \times 10^0$, for N×C, and also $[T_{MA}] \times 10^1 + [U_{MA}] \times 10^0$, for M×A, $[T_{MB}] \times 10^1 + [U_{MB}] \times 10^0$, for M×B, and $[T_{MC}] \times 10^1 + [U_{MC}] \times 10^0$, for M×C, we have:

Mental Mathematics

$\Rightarrow N \times (ABC) = [N \times A] \times 10^2 + [N \times B] \times 10^1 + [N \times C] \times 10^0$

As N is a digit (0 - 9), and each of A, B, or C is also a digit, thus each of the N×A, N×B, and N×C is also a number with two digits at most.

Note: Each of the N×A, N×B, and N×C lies between zero to eighty-one.

Now by using symbols, **U**, and **T** as **Units Digit** and **Tens Digit**, we may write, each of the N×A, N×B, and N×C, as follow:

$N \times A = [T_A] \times 10^1 + [U_A] \times 10^0$

$N \times B = [T_B] \times 10^1 + [U_B] \times 10^0$

$N \times C = [T_C] \times 10^1 + [U_C] \times 10^0$

Now by replacing, $[T_A] \times 10^1 + [U_A] \times 10^0$, for N×A, and $[T_B] \times 10^1 + [U_B] \times 10^0$, for N×B, and $[T_C] \times 10^1 + [U_C] \times 10^0$, for N×C, in $[N \times A] \times 10^2 + [N \times B] \times 10^1 + [N \times C] \times 10^0$, we obtain:

$[N \times A] \times 10^2 + [N \times B] \times 10^1 + [N \times C] \times 10^0 =$

$[(T_A)) \times 10^1 + (U_A) \times 10^0] \times 10^2 + [(T_B) \times 10^1 + (U_B) \times 10^0] \times 10^1 + [(T_C) \times 10^1 + (U_C) \times 10^0] \times 10^0$

$\Rightarrow N \times (ABC) = (T_A) \times 10^3 + (U_A) \times 10^2 + (T_B) \times 10^2 + (U_B) \times 10^1 + (T_C) \times 10^1 + (U_C) \times 10^0$

Now we will do factorization of the terms, with the same factor:

$\Rightarrow N \times (ABC) = (T_A) \times 10^3 + [(U_A) + (T_B)] \times 10^2 + [(U_B) + (T_C)] \times 10^1 + (U_C) \times 10^0$

$\Rightarrow \mathbf{N \times (ABC) = [T_A][(U_A) + (T_B)] [(U_B) + (T_C)][U_C]}$

These formulas prove the correctness of the idea, we explained in chapter 2.

Example 7 – 8

What is the product of 287×4?

Mental Mathematics

Answer:

⇨ ABC = 287, and N = 4 ⇨ A = 2, B = 8, and C = 7.

4×7 = 28 thus U_C = **8**, and **Tc = 2**

4×8 = 32, thus U_B = **2**, and T_B = **3**.

4×2 = 08, thus U_A = **8**, and T_A = **0**.

⇨ $[T_A][(U_A) + (T_B)] [(U_B) + (T_C)][U_C]$ = 0[8 + 3][2 + 2][8] = 0[11]58 = [0 + 1]158 = **1148**.

Multi Digit Numbers Multiplication – Units Digit and Tens Digit Method

Now we will multiply a three digit number ABC, by a two digit number MN.

$(ABC) \times (MN) = [(A \times 10^2 + B \times 10^1 + C \times 10^0)] \times [M \times 10^1 + N \times 10^0]$

$= (M \times A) \times 10^3 + (M \times B) \times 10^2 + (M \times C) \times 10^1 + (N \times A) \times 10^2 + (N \times B) \times 10^1 + (N \times C) \times 10^0$

As each of N, and M is a digit, and each of A, B, or C is also a digit, thus each of the N×A, N×B, N×C, M×A, M× B, and M×C is a two digit number too (0 – 81).

Now by using symbols, **U**, and **T** as **Units Digit** and **Tens Digit**, we may write, each of the N×A, N×B, N×C, M×A, M× B, and M×C, as follow:

N×A = $[T_{NA}] \times 10^1 + [U_{NA}] \times 10^0$

N×B = $[T_{NB}] \times 10^1 + [U_{NB}] \times 10^0$

N×C = $[T_{NC}] \times 10^1 + [U_{NC}] \times 10^0$

M×A = $[T_{MA}] \times 10^1 + [U_{MA}] \times 10^0$

M×B = $[T_{MB}] \times 10^1 + [U_{MB}] \times 10^0$

M×C = $[T_{MC}] \times 10^1 + [U_{MC}] \times 10^0$

Now by plugging, $[T_{NA}] \times 10^1 + [U_{NA}] \times 10^0$, for N×A, and $[T_{NB}] \times 10^1 + [U_{NB}] \times 10^0$, for N×B, and $[T_{NC}] \times 10^1 + [U_{NC}] \times 10^0$, for N×C, and also, $[T_{MA}] \times 10^1 + [U_{MA}] \times 10^0$, for M×A, $[T_{MB}] \times 10^1 + [U_{MB}] \times 10^0$, for M×B, and $[T_{MC}] \times 10^1 + [U_{MC}] \times 10^0$, for M×C, we have:

Mental Mathematics

$(M \times A) \times 10^3 + (M \times B) \times 10^2 + (M \times C) \times 10^1 + (N \times A) \times 10^2 + (N \times B) \times 10^1 + (N \times C) \times 10^0 =$

$\{[T_{MA}] \times 10^1 + [U_{MA}] \times 10^0\} \times 10^3 + \{[T_{MB}] \times 10^1 + [U_{MB}] \times 10^0\} \times 10^2 + \{[T_{MC}] \times 10^1 + [U_{MC}] \times 10^0\} \times 10^1 + \{[T_{NA}] \times 10^1 + [U_{NA}] \times 10^0\} \times 10^2 + \{[T_{NB}] \times 10^1 + [U_{NB}] \times 10^0\} \times 10^1 + \{[T_{NC}] \times 10^1 + [U_{NC}] \times 10^0\} \times 10^0$

$= [T_{MA}] \times 10^4 + [U_{MA}] \times 10^3 + [T_{MB}] \times 10^3 + [U_{MB}] \times 10^{2\,+} \{[T_{MC}] \times 10^2 + [U_{MC}] \times 10^1 + [T_{NA}] \times 10^3 + [U_{NA}] \times 10^2 + [T_{NB}] \times 10^2 + [U_{NB}] \times 10^1 + [T_{NC}] \times 10^1 + [U_{NC}] \times 10^0$

Now we will put all terms with the same factor, as one group, by factorization as follow:

$[T_{MA}] \times 10^4 + [U_{MA} + T_{MB} + T_{NA}] \times 10^3 + [U_{MB} + T_{MC} + U_{NA}\, T_{NB}] \times 10^2 + [U_{MC} + U_{NB} + T_{NC}] \times 10^1 + [U_{NC}] \times 10^0$

$= [T_{MA}][U_{MA} + T_{MB} + T_{NA}][U_{MB} + T_{MC} + U_{NA}\, T_{NB}][U_{MC} + U_{NB} + T_{NC}][U_{NC}]$

⇨ **MN×ABC** $= [T_{MA}][U_{MA} + T_{MB} + T_{NA}][U_{MB} + T_{MC} + U_{NA}\, T_{NB}][U_{MC} + U_{NB} + T_{NC}][U_{NC}]$

Example – (7 – 9)

What is the product of 37 by 253?

37×253 = **MN×ABC** $= [T_{MA}][U_{MA} + T_{MB} + T_{NA}][U_{MB} + T_{MC} + U_{NA}\, T_{NB}][U_{MC} + U_{NB} + T_{NC}][U_{NC}]$

M=3, N = 7, A = 2, B = 5, and C = 3.

M×A = 3×2 = 06, thus T_{MA} = 0, and U_{MA} = 6.

M×B = 3×5 = 15, thus T_{MB} = 1, and U_{MB} = 5.

M×C = 3×3 = 09, thus T_{MC} = 0, and U_{MC} = 9.

N×A = 7×2 = 14, thus T_{NA} = 1, and U_{NA} = 4.

N×B = 7×5 = 35, thus T_{NB} = 3, and U_{NB} = 5.

N×C = 7×3 = 21, thus T_{NC} = 2, and U_{NC} = 1.

⇨ $[T_{MA}][U_{MA} + T_{MB} + T_{NA}][U_{MB} + T_{MC} + U_{NA} + T_{NB}] \times 10^2][U_{MC} + U_{NB} + T_{NC}][U_{NC}] =$

Mental Mathematics

⇨ 37×253 = [0][1 + 1 + 6][3 + 4 + 0 + 5][9 + 5 + 2][1] = [0][8][12][16]1 = 09361.

Mental Mathematics

$(M \times A) \times 10^3 + (M \times B) \times 10^2 + (M \times C) \times 10^1 + (N \times A) \times 10^2 + (N \times B) \times 10^1 + (N \times C) \times 10^0 =$

$\{[T_{MA}] \times 10^1 + [U_{MA}] \times 10^0\} \times 10^3 + \{[T_{MB}] \times 10^1 + [U_{MB}] \times 10^0\} \times 10^2 + \{[T_{MC}] \times 10^1 + [U_{MC}] \times 10^0\} \times 10^1 + \{[T_{NA}] \times 10^1 + [U_{NA}] \times 10^0\} \times 10^2 + \{[T_{NB}] \times 10^1 + [U_{NB}] \times 10^0\} \times 10^1 + \{[T_{NC}] \times 10^1 + [U_{NC}] \times 10^0\} \times 10^0$

$= [T_{MA}] \times 10^4 + [U_{MA}] \times 10^3 + [T_{MB}] \times 10^3 + [U_{MB}] \times 10^2 + \{[T_{MC}] \times 10^2 + [U_{MC}] \times 10^1 + [T_{NA}] \times 10^3 + [U_{NA}] \times 10^2 + [T_{NB}] \times 10^2 + [U_{NB}] \times 10^1 + [T_{NC}] \times 10^1 + [U_{NC}] \times 10^0$

Now we will put all terms with the same factor, as one group, by factorization as follow:

$[T_{MA}] \times 10^4 + [U_{MA} + T_{MB} + T_{NA}] \times 10^3 + [U_{MB} + T_{MC} + U_{NA} T_{NB}] \times 10^2 + [U_{MC} + U_{NB} + T_{NC}] \times 10^1 + [U_{NC}] \times 10^0$

$= [T_{MA}][U_{MA} + T_{MB} + T_{NA}][U_{MB} + T_{MC} + U_{NA} T_{NB}][U_{MC} + U_{NB} + T_{NC}][U_{NC}]$

⇨ **MN×ABC** = $[T_{MA}][U_{MA} + T_{MB} + T_{NA}][U_{MB} + T_{MC} + U_{NA} T_{NB}][U_{MC} + U_{NB} + T_{NC}][U_{NC}]$

Example – (7 – 9)

What is the product of 37 by 253?

37×253 = **MN×ABC** = $[T_{MA}][U_{MA} + T_{MB} + T_{NA}][U_{MB} + T_{MC} + U_{NA} T_{NB}][U_{MC} + U_{NB} + T_{NC}][U_{NC}]$

M=3, N = 7, A = 2, B = 5, and C = 3.

M×A = 3×2 = 06, thus **T_{MA}** = 0, and **U_{MA}** = 6.

M×B = 3×5 = 15, thus **T_{MB}** = 1, and **U_{MB}** = 5.

M×C = 3×3 = 09, thus **T_{MC}** = 0, and **U_{MC}** = 9.

N×A = 7×2 = 14, thus **T_{NA}** = 1, and **U_{NA}** = 4.

N×B = 7×5 = 35, thus **T_{NB}** = 3, and **U_{NB}** = 5.

N×C = 7×3 = 21, thus **T_{NC}** = 2, and **U_{NC}** = 1.

⇨ $[T_{MA}][U_{MA} + T_{MB} + T_{NA}][U_{MB} + T_{MC} + U_{NA} + T_{NB}] \times 10^2][U_{MC} + U_{NB} + T_{NC}][U_{NC}]$ =

Mental Mathematics

⇨ 37×253 = [0][1 + 1 + 6][3 + 4 + 0 + 5][9 + 5 + 2][1] = [0][8][12][16]1 = 09361.

Mental Mathematics

Mental Mathematics

Mental Mathematics

www.ingramcontent.com/pod-product-compliance
Lightning Source LLC
Chambersburg PA
CBHW082206220526
45470CB00010B/3067